# DATE DUE

| | | | |
|---|---|---|---|
| | | | |
| | | | |
| | | | |
| | | | |
| | | | |
| | | | |
| | | | |
| | | | |
| | | | |
| | | | |
| | | | |
| | | | |
| | | | |
| | | | |
| | | | |
| | | | |
| | | | |
| | | | |

# ELECTRON
# TRANSFER REACTIONS OF
# COMPLEX IONS IN SOLUTION

# CURRENT CHEMICAL CONCEPTS

A Series of Monographs

Editor: ERNEST M. LOEBL

R. McWEENY: Spins in Chemistry, 1970

Edited by LOUIS MEITES

HENRY TAUBE: Electron Transfer Reactions of Complex Ions in Solution, 1970

P. ZUMAN: The Elucidation of Organic Electrode Processes, 1969

FRANK A. BOVEY: Polymer Conformation and Configuration, 1969

H. J. EMELÉUS: The Chemistry of Fluorine and Its Compounds, 1969

# Electron
# Transfer Reactions of
# Complex Ions in Solution

HENRY TAUBE
DEPARTMENT OF CHEMISTRY
STANFORD UNIVERSITY
STANFORD, CALIFORNIA

1970

ACADEMIC PRESS    New York and London

ACADEMIC PRESS, INC.
111 Fifth Avenue, New York, New York 10003

*United Kingdom Edition published by*
ACADEMIC PRESS, INC. (LONDON) LTD.
Berkeley Square House, London W1X 6BA

LIBRARY OF CONGRESS CATALOG CARD NUMBER: 71-113742

PRINTED IN THE UNITED STATES OF AMERICA

# FOREWORD

This is one of a series of monographs made possible by a Science Center Development Grant from the National Science Foundation to the Polytechnic Institute of Brooklyn. That grant enabled the Department of Chemistry to establish a Distinguished Visiting Lectureship that is held successively by a number of eminent chemists, each of whom has played a leading part in the development of some important area of chemical research. During his term of residence at the Institute, each lecturer gives a series of public lectures on a topic of his choice.

These monographs arose from a desire to preserve the substance of these lectures and to share them with interested chemists everywhere. They are intended to be more leisurely, more speculative, and more personal than reviews that might have been published in other ways. Each of them sets forth an outstanding chemist's views on the past, the present, and the possible future of his field. By showing how the facts of yesterday have given rise to today's concepts, deductions, hopes, fears, and guesses, they should serve as guides to the research and thinking of tomorrow.

This volume is an expanded and updated treatment of the material presented in a series of four lectures by Professor

v

Taube while in residence at the Institute in November and
December, 1967. It is with great pride and pleasure that we
present this record of the stimulation and profit that our department obtained from his visit.

Louis Meites, Editor                    F. Marshall Beringer

*Head, Department of Chemistry*

# CONTENTS

*Foreword*                                                                    v


Chapter I. INTRODUCTION OF THE REAGENTS: STUDIES IN
              HYDRATION OF CATIONS

   I. Introduction                                                1
  II. Summary of Results                                            3
 III. Application of the Methods                                       5
 IV. Relations between Substitution Rates                             13
  V. Comments on Rates Summarized in Table I-1                    20
 VI. Problems                                                         22

    References                                          24


Chapter II. DESCRIPTION OF THE ACTIVATED COMPLEXES FOR
               ELECTRON TRANSFER

   I. Introduction                                               27
  II. The Outer-Sphere Activated Complex                           28
 III. The Inner-Sphere Activated Complex                              35

    References                                          45

Chapter III.　Some Aspects of Ligand Effects in Electron Transfer Reactions

| | |
|---|---|
| I. Introduction | 49 |
| II. Simple Ligands | 51 |
| III. Activation Parameters | 59 |
| IV. The Mechanism of Electron Transfer | 62 |
| V. Conclusion | 68 |
| References | 71 |

Chapter IV.　Induced Electron Transfer

| | |
|---|---|
| I. Introduction | 73 |
| II. History | 76 |
| III. Evidence for the Formation of an Intermediate | 78 |
| IV. The Oxidation of Pentaamminepyridinemethanolcobalt (III) by Ce(IV) | 80 |
| V. Low-Spin Co(II) as an Intermediate | 90 |
| VI. Other Systems | 94 |
| References | 97 |

| | |
|---|---|
| Subject Index | 99 |

# I

---

# INTRODUCTION OF THE REAGENTS:
## Studies in Hydration of Cations

### I. INTRODUCTION

Questions of mechanisms of redox reactions plainly involve the disposition of groups in the first coordination spheres of both partners, before, during, and after the change in oxidation state takes place. The problem of "during" is the one peculiar to a discussion of mechanism. It can be only discussed meaningfully when we know what the changes in the coordination of the partners are corresponding to the net changes. Among common redox couples are those exemplified by the half-reaction

$$SO_2(aq) + 2H_2O = 3H^+ + HSO_4^- + 2e^- \qquad (I-1)$$

in which changes in the first coordination sphere are easily recognizable, as well as those exemplified by the $Fe(CN)_6^{4-,3-}$ half-reaction

$$Fe(CN)_6^{4-} = Fe(CN)_6^{3-} + e^- \qquad (I-2)$$

in which it is known with equal certainty that the first coordination spheres of the ions remain · unaltered in composition on

1

electron transfer. But among the important couples are others, such as

$$Fe^{2+}(aq) = Fe^{3+}(aq) + e^-  \qquad (I\text{-}3)$$

and

$$Fe^{2+}(aq) + HF = FeF^{2+}(aq) + H^+ + e^-  \qquad (I\text{-}4)$$

Here the problem of the state of coordination with respect to water of components of the couple must be settled before satisfactory progress in understanding the mechanism of electron transfer can be expected. A study of the hydration (or solvation) of ions is, therefore, a necessary preliminary to the study of the mechanism of many redox reactions. Though much progress in understanding ion hydration has been made, important questions remain to be answered, and the subject of ion hydration, let alone the more general one of ion solvation, remains a live research area.

In my efforts to understand the mechanism of redox reactions, an interest in ion hydration preceded and was a preparation for experiments which were planned in the hope of illuminating redox mechanisms. For the reasons mentioned, and because the discussion of ion hydration is also a means of introducing properties of the various couples which are necessary for understanding the strategy used in the study of redox mechanisms, I am beginning with a discussion of ionic hydration.

In dealing with ionic hydration in the present context, a knowledge of the labilities of the coordination spheres is as important as the knowledge of composition and structure. The labilities of the reacting complexes impose limitations on the mechanisms of the electron-transfer reactions. Moreover, much of the progress in defining the geometries of the activated complexes has been made by exploiting information about the labilities of reactants and products. The discussion of hydration will, therefore, be concerned not only with the compositions and structures of the aquo complexes, but also with their labilities. The discussion will, nevertheless, be severely restricted, emphasizing those methods which provide direct information on the three aspects mentioned as applied to ions which have been fairly intensively studied as reactants or products in redox processes.

## II. SUMMARY OF RESULTS

The two principal methods which have been used for determining the compositions of the first coordination spheres of the aquo ions with which we will be concerned are isotopic dilution using labeled oxygen, and electronic absorption spectroscopy, the latter being the more widely useful. Nuclear magnetic resonance spectroscopy, using [17]O or H, is an important method for the general problem of solvation or hydration, but has been of limited utility for the particular subject area which has been defined. In measuring labilities, isotopic dilution and NMR ([17]O and H) have been used, the NMR method being particularly powerful. As will be shown, rate comparisons also are a guide in getting at least an approximate idea of the rate of a particular substitution reaction.

The results of the studies devoted to learning the formulas and labilities of some aquo ions of particular interest for a discussion of redox reactions are shown in Table I-1. The table is also intended as a compendium of information on other properties of vital interest—namely, the values of $E^0$ for various couples—and so it includes couples which have not been completely characterized with respect to the compositions and labilities of the aquo complexes.

It will be noted that all the aquo ions shown in Table I-1 for which the formulas have been determined have a coordination number of six. The structures of those having coordination numbers of six almost always correspond to regular octahedra. The ions $Cr^{2+}$ and $Cu^{2+}$ are unusual in that the coordination octahedra are strongly distorted, and it is assumed that the two water molecules along one axis are much farther from the cation than the four in equatorial positions. The coordination number of

TABLE I-1

PROPERTIES OF SOME REDOX COUPLES[a]

| Reduced form | Oxidized form | $E^0$ [1] |
|---|---|---|
| $Eu(aq)^{2+}$ <br> very rapid | $Eu(aq)^{3+}$ <br> very rapid | —0.43 |
| $Cr(H_2O)_4(H_2O')_2^{2+}$ <br> $> 10^9$ [2] | $Cr(H_2O)_6^{3+}$ <br> $3 \times 10^{-6}$ [3, 4] | —0.41 |
| $V(H_2O)_6^{2+}$ <br> $1.0 \times 10^2$ [5] | $V(H_2O)_6^{3+}$ <br> $\sim 10^2$ [6] | —0.26 |
| $Cu^+(aq)$ <br> very rapid | $Cu(H_2O)_4(H_2O')_2^{2+}$ <br> $\sim 10^9$ [2] | 0.15 |
| $Ru(NH_3)_5(H_2O)^{2+}$ <br> $\sim 10$ [8] | $Ru(NH_3)_5(H_2O)^{3+}$ <br> $< 10^{-6}$ [9] | 0.16 [7] |
| $Co(H_2O)_6^{2+}$ <br> $2 \times 10^5$ [10] | $Co(NH_3)_5(H_2O)^{3+}$ <br> $6 \times 10^{-6}$ [11, 12] | 0.3 [13] |
| $Fe(H_2O)_6^{2+}$ <br> $10^6$ [10] | $Fe(H_2O)_6^{3+}$ <br> $1-4 \times 10^2$ [14] | 0.77 |
| $Ce^{3+}(aq)$ <br> very rapid | $Ce(IV)$ [15] <br> very rapid | 1.61 |
| $Co(H_2O)_6^{2+}$ <br> $2 \times 10^5$ [10] | $Co(H_2O)_6^{3+}$ <br> $1 \times 10^2$ [6] | 1.8 |

[a] Values of the specific rate constant $(sec^{-1})$ for the exchange of water between coordination sphere and solvent are given below formulas.

six is not to be regarded as a rule of nature, however, for it is entirely likely that the coordination numbers for $Cu^+$, $Eu^{2+}$, $Ce^{3+}$, $Eu^{3+}$, and $Ce(IV)$ will prove to be different from six when they are determined. The prevalence of six in Table I-1 is a result of the fact that most of the species are transition metal ions of charge 2 and 3, and these ions have radii appropriate for coordination number six. In the matter of labilities, the large differences between ions of the same charge ($Co^{2+}$, $2 \times 10^5$ vs. $Cu^{2+}$, $>10^9$; $Cr^{3+}$, $3 \times 10^{-6}$ vs. $Fe^{3+}$, $\sim 10^2$), and the differences among ions of the same element in different oxidation states ($Cr^{2+}$, $>10^9$ vs. $Cr^{3+}$, $3 \times 10^{-6}$; $Fe^{2+}$, $10^6$ vs. $Fe^{3+}$, $\sim 10^2$) are to be noted, as is the difference between water in $Co(H_2O)_6^{3+}$ (probably 1 to 100) compared with $Co(NH_3)_5OH_2^{3+}$ ($6 \times 10^{-6}$). The rates of substitution for some of the ions, namely for $Cr(H_2O)_6^{3+}$, $Ru(NH_3)_5H_2O^{3+}$, and $Co(NH_3)_5H_2O^{3+}$, are slow enough so that the time scale for the change is of the order of days. This point will take on added significance when the chemistry of the redox processes for the corresponding couples is described.

## III. APPLICATION OF THE METHODS

### A. *Analysis of Absorption Spectra*

This is by far the most widely useful of the methods which are applicable to determining the state of coordination of the cations. The power of the method is illustrated in Fig. I-1 for $Co^{2+}$(aq) by results taken from papers of Schmitz-Dumont and co-workers [16]. Measurements of the absorption spectrum of CoO in MgO, in which the dipositive ions are known to occupy octahedral holes, and in $MgAl_2O_4$, in which the dipositive ions occupy tetrahedral holes, demonstrate the sensitivity of the spectrum to the coordination number of the ion. The similarity of the absorption spectrum of $Co^{2+}$(aq) to the former spectrum—and at the same time its dissimilarity to the latter—shows that the dominant form of the aquo complex in solution has the form $Co(H_2O)_6^{2+}$. Conclusions of this type have a good theoretical foundation in most cases, and the theory applied to the analysis of the spectra is powerful enough to support the conclusion that

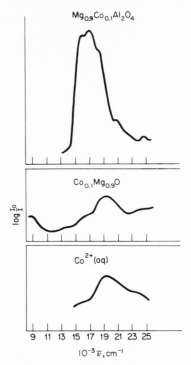

FIG. I-1.   The absorption spectra of $Co^{2+}$ in different environments (from [16]).

the coordination spheres of $Cu^{2+}$(aq) and $Cr^{2+}$ are strongly distorted. Though in some cases the observations on the absorption spectra have not explicitly been put to this particular use, they prove (or at least support) the conclusions embodied in the formulas entered in Table I-1 for the aquo ions of Cr(II), Cr(III), V(II), V(III), Cu(II), Co(II), Fe(II), Fe(III), and Co(III).

## B. *Isotopic Dilution*

Though the approach to ion solvation based on the analysis of absorption spectra has wide application, it by no means encompasses all the ions of interest, and has in any case little to say about labilities. Perhaps it is precisely because the isotopic dilution method, despite the restricted field of ions to which it can be applied, can at once give information about composition and lability, that it played an important role in the modern development of the subject of ionic hydration. The first cation–water complex to be described [3] with respect to formula and rate of exchange of water between the aquo complex and the solvent was $Cr(H_2O)_6^{3+}$, and the conclusions were based on isotopic dilution studies using $^{18}O$ as tracer. As will be appear in the following, the scope of the method, if extreme conditions are of interest, is really quite wide; it will also appear, however, that the method is tedious and difficult except in special cases.

The principle of isotopic dilution as applied to ionic solvation is identical with that of its use in quantitative analysis. The analytical problem in the present application is to determine the amount of water bound to an ion such as $Cr^{3+}$ when the total amount of water in the solution containing this species is known. A distinction between water bound to the cation and free water is possible when there is a delay in the dilution by the bound water of an isotopically enriched water sample which is added to the solution. If a sample of water can be removed, say by distillation or solvent extraction, before the isotopically enriched water is diluted by the bound water, the amount of water "held back" by the dissolved salt can be determined. By removing samples of water at intervals to assay the isotopic composition, the rate of exchange between bound water and solvent can be measured.

The results of an experiment reported in the first application of isotopic dilution to the study of ion hydration are reproduced in Table I-2. For present purposes, the data are chiefly useful in assessing the sensitivity of the method. Even in the early work, it proved to be possible to determine the isotopic composition of the solvent (this was separated by evaporation) to an accuracy of $\pm 2$ parts per thousand. At the concentration level in the experiment used for illustration, a holdback of six molecules of water

TABLE I-2

EXCHANGE OF $Cr(aq)^{3+}$ WITH $H_2O$

| Experimental | |
| --- | --- |
| Time, hr | $N^a$ |
| 0.30 | 8.422 |
| 18.3 | 8.169 |
| 18.9 | 8.169 |
| 401.3 | 7.518 |

Calculated

| | | | |
| --- | --- | --- | --- |
| $N_{0,exp}$ : | 8.427 | $N_{0,exp}/N_{\infty,exp}$ : | 1.1211 |
| $N_0^5$ : | 8.307 | $N_0^6/N_{\infty,calc}$ : | 1.1177 |
| $N_0^6$ : | 8.474 | $N_0^7/N_{\infty,calc}$ : | 1.1412 |
| $N_0^7$ : | 8.653 | $N_0^6/N_{0,exp}$ : | 1.0056 |
| $N_{\infty,exp}$ : | 7.517 | $N_{\infty,calc}/N_{\infty,exp}$ : | 1.0087 |
| $N_{\infty,calc}$ : | 7.582 | | |

[a] $N$ expresses the mole fraction $H_2^{18}O/(H_2^{18}O + H_2^{16}O)$ in the samples distilled from the solutions. From these, values of $N$ for zero and infinite time ($N_{0,exp}$ and $N_{\infty,exp}$, respectively) are obtained by extrapolation. $N_0^5$, $N_0^6$, and $N_0^7$ are the calculated values of the mole fraction in the liquid for initial holdback by each $Cr^{3+}$ of five, six, and seven molecules of $H_2O$, respectively. $N_{\infty,calc}$ is the mole fraction calculated for random mixing in the liquid. The solutions contained 1.57 molal $Cr(ClO_4)_3$ and 0.13 molal $HClO_4$, and the data were obtained at 25°.

per chromic ion compared to seven changes the isotopic composition of the solvent by three parts per hundred so that the holdback can be determined with considerable accuracy. The ratio of 1.005 for $N_0^6/N_0$ is no cause for concern, because it is almost exactly the isotopic fractionation for the evaporation of water under the conditions of the experiment.

The method described is limited by the difficulty of sampling the solvent immediately after mixing enriched water with the solution, and is applicable only to ions for which the time needed for substantial exchange is longer than the time between mixing and sampling. By using a flow adaption of the method, it is possible greatly to shorten the critical time period. An attempt [17] to develop such a method was made using the apparatus shown in Fig. I-2. The separate solutions are brought together in a mixing chamber and sampling is done by evaporating from the flowing stream as it issues from the delivery tube. The

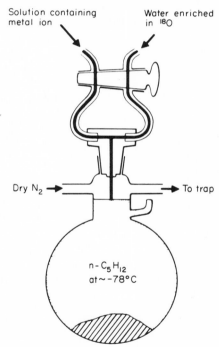

Solution containing
metal ion

Water enriched
in $^{18}O$

Dry $N_2$ →

→ To trap

$n\text{-}C_5H_{12}$
at ~ -78°C

FIG. I-2.   Apparatus for study of solvation by the isotopic dilution method adapted to rapid measuring and sampling.

exchange reaction is quenched when the stream impinges on the surface of a chilled liquid of lower density ($C_5H_{12}$ was used). By the flow technique, the time between mixing and sampling can be reduced to less than 0.01 sec. The method, though it appears to have great potential for non-aqueous solvents which can be cooled to low temperatures, has not been sufficiently studied or refined. In the evaporation from the flowing stream, large isotopic fractionation factors were encountered, and these produced effects almost as large as those arising from the actual holdback. Even with this limitation, significant results were obtained by using a solution of chromic ion as a calibrating liquid. It was shown that the holdback of water by $Al^{3+}$ is identical with that by $Cr^{3+}$, thus justifying the formula $Al(H_2O)_6^{3+}$ for the hydrated ion, and that the half-time for water exchange on $Al(H_2O)_6^{3+}$ is greater than 0.01 sec. Both conclusions have since been corroborated by other means [18]. It was also shown that $Fe^{3+}(aq)$ and $Ni^{2+}(aq)$ exchange water bound in the first coordination sphere in less than 0.02 sec, and this conclusion is consistent with measurements of the lability which have been made using the line-broadening technique to be described in the next section.

## C. *Nuclear Magnetic Resonance Measurements*

When the exchange of water between hydrated ions and solvent is slow enough, a separate NMR signal for $^{17}O$ in the coordination sphere of the ion from that of $^{17}O$ in the bulk solvent can be expected. When separate signals are observed, a convenient method of measuring the amount of water in the first coordination sphere of the ion is at hand. This method, direct though it is in principle, has not been especially useful for ions commonly used as redox reagents. Such ions usually are paramagnetic, and the $^{17}O$ signal for $H_2O$ in the first coordination sphere is broadened to such an extent that the bound water has been observed directly for only one ion, namely $Ni^{2+}$ [19]. Even in this case, the absorption band characteristic of the bound water is barely detectable and provides no basis for a dependable determination of coordination number, at least by the integration of absorption intensity. This method has, however, yielded definitive conclusions for diamagnetic ions, but here the chemical shift for the bound water is so small that the bulk solvent and coordinated water peaks overlap. However, data of high quality have been obtained by taking advantage of the fact that when certain paramagnetic and substitution-labile metal ions are added, they shift and broaden the NMR absorption for $^{17}O$ contained in the solvent much more than the absorption for that contained in the coordination sphere of the ion under investigation. Two methods of determining coordination numbers are then available [19]: the direct one of integrating the absorption and the less direct "molal shift" method. The principle of the latter method is that the shift in bulk water absorption produced by the substitution-labile ion added at a given temperature depends to a first approximation only on the ratio of this metal ion to free water, and thus the shift can be used to determine the free water.

At low temperatures, the exchange of protons between coordinated water and solvent may be slow enough that proton absorption can be used in much the same way as $^{17}O$ absorption, often without resort to the paramagnetic shift method.

For the particular class of ions we are concerned with, the NMR method has been particularly useful in determining labilities, and of the two NMR possibilities for water, namely using $^{17}O$ or H, the former is more widely applicable at room temperature (the H—O bond is usually—though not always—more labile than the metal ion—O bond, and in the usual case the rate of H exchange sets only an upper limit on the rate of water exchange). In measuring the broadening of the $^{17}O$ absorption in water caused by a paramagnetic ion, three different temperature regions generated by two parallel paths can, in general, be expected. At a high enough temperature, when substitution in the first sphere of coordination is fast enough, the path involving the first sphere of coordination is dominant;

at the high end of the high-temperature range, substitution can be rapid enough that the interaction between electrons and the nuclear spins determines the rate of relaxation of the nuclear spin. The rate by this path increases as the temperature is lowered, but since this mechanism depends on a fresh supply of nuclei being always provided, as the temperature is lowered, the rate at which the nuclei are exposed to the influence of the electron spins can become rate-determining. When this situation obtains, because the rate of exchange decreases as the temperature is lowered, eventually relaxation by some parallel path, such as by outer-sphere interactions, will become dominant. Here it is unlikely that substitution will be rate-determining, but rather the interaction of the nuclear spin with the perturbing influence, and the relaxation rate will again increase as the temperature decreases. All three regions have not been observed for any one

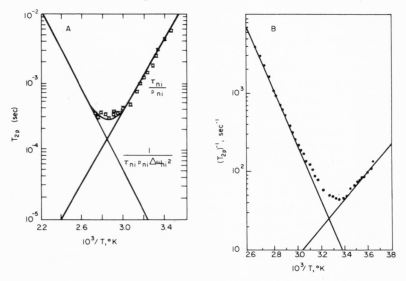

FIG. I-3.   $O^{17}$ line-broadening by $Ni^{2+}(aq)$ (A, from [10]) and by $V^{2+}(aq)$ (B, from [5]).

ion. The observations for $Ni(H_2O)_6^{2+}$ and $V(H_2O)_6^{2+}$ together do show up three regions, and are reproduced in Fig. I-3. The region of exchange-controlled relaxation can be recognized as showing $1/T_{2p}$ [10] to decrease with the temperature. As already mentioned, for the other two regions, $1/T_{2p}$ increases as the temperature decreases.

For the ions $Mn^{2+}$, $Co^{2+}$, $Fe^{2+}$, and $Fe^{3+}$, as well as for the two already mentioned, the NMR line-broadening method as applied to determining the labilities of the coordination spheres is reasonably straightforward [10]. For $Cu^{2+}$ and $Cr^{2+}$, the situation is more complex [2, 10], but it is at the same time particularly interesting.

These ions are believed to have coordination shells derived from the octahedron by an axial distortion, and in both cases probably by lengthening the metal ion$-H_2O$ distance along the axis. The analysis of the NMR data is consistent [2] with lower limits of $1 \times 10^9$ ($Cu^{2+}$) and $7 \times 10^9$ sec$^{-1}$($Cr^{2+}$) for the specific rates of the exchange process

$$(H_2O)_a + (H_2O)_l = (H_2O)_a + (H_2O)_l \qquad (I-5)$$

where the subscript a refers to water in the labile axial position of the complex, and 1 to the liquid, and with the conclusion that the limits quoted measure the rate of the pseudorotation in which axial water becomes equatorial.

## IV. RELATIONS BETWEEN SUBSTITUTION RATES

The replacement of water by water is only one of the many kinds of substitution reactions of metal ions which it is important to consider in pursuing an interest in the mechanisms of redox reactions. When dealing with reactions in water as solvent, there is no need to apologize for a special interest in the replacement of water by water on a metal ion, but it is nevertheless important to consider whether the rates of these reactions are a guide to those in which some ligand other than water replaces water molecules in the first coordination sphere, or to those in which neither the entering group nor the leaving one is a water molecule.

A striking relationship between the rate of replacement of water by water compared to that by another ligand has emerged from the study of ligand replacement for the labile metal centers [21]. This should not be taken to imply that the relationship is limited to these ions; the connection was first recognized for the labile ions because very extensive data covering many different ligands and metal ions were accumulated for them by the application of new techniques for measuring the rates of rapid reactions. The relationship referred to is that the rates of the processes

$$H_2O_h \cdot H_2O_l^* = H_2O_h^* \cdot H_2O_l \qquad (I\text{-}6)$$

$$H_2O_h \cdot L = L_h \cdot H_2O_l \qquad (I\text{-}7)$$

where the subscripts h and l refer to the hydration sphere and the liquid, respectively, and where $H_2O_h \cdot L$ represents the outer-sphere complex of L, are nearly the same. The general agreement of rates of substitution when expressed in this way is most convincingly demonstrated for the sulfate complexes of the dipositive ions, because for these the stabilities of the outer-sphere complexes are most accurately known. In Table I-3 the data on rates of water–water interchange and water–sulfate–ion interchange are set forth.

If the entries for $V(H_2O)_6^{2+}$ are omitted, and this is probably justified because the complete data for the $V(H_2O)_6^{2+} \cdot SO_4^{2-}$

TABLE I-3

COMPARISON OF WATER–WATER AND SULFATE–WATER
INTERCHANGE RATES FOR DIPOSITIVE IONS[a] [22]

|  | $V^{2+}$ | $Mn^{2+}$ | $Fe^{2+}$ | $Co^{2+}$ | $Ni^{2+}$ |
|---|---|---|---|---|---|
| $H_2O_h \cdot H_2O_l$ | $1.0 \times 10^2$ | $3 \times 10^7$ | $3 \times 10^6$ | $1.0 \times 10^6$ | $2.5 \times 10^4$ |
| $H_2O_h \cdot SO_4^{2-}$ | $1 \times 10^3$ | $4 \times 10^6$ | $1.0 \times 10^6$ | $2.0 \times 10^5$ | $1.6 \times 10^4$ |

[a] In $sec^{-1}$.

system have not been published, the agreement between the rates for the two kinds of reaction is seen to be remarkably close. It is tempting on the basis of agreement of the kind noted, particularly when it extends to other ligands and metal centers as well, to conclude that the reactions are $S_N1$ in character. Such a conclusion, reasonable though it is, when it is based only on rate comparisons of the kind illustrated, must be regarded as being premature and insecure. In the first place, the agreement shown in Table I-3 is not exact, nor would exact agreement be expected even if the reactions were purely $S_N1$ in character. An intermediate such as $Ni(H_2O)_5^{2+}$ can be expected to exert some discrimination between groups immediately adjacent to it. Differences are expected whether reaction is by an $S_N1$ or $S_N2$ process, and it is difficult to decide whether those observed are of the magnitude expected for an $S_N1$ process or for an $S_N2$ process. This is particularly true because the mechanism of substitution has not been settled unequivocally for an octahedral $2+$ ion, and there is no basis for deciding how large the differences in rate should be as the ligand changes either for an $S_N1$ or $S_N2$ mechanism.

Though it is acknowledged that by now more extensive data exist for rates of substitution on labile metal ions—more extensive at least with respect to variation in the nature of the metal ion— the problem of mechanism has been much more thoroughly investigated for the substitution-inert metal ions, and this experience has provided lessons which are applicable also to the more labile systems. It should be emphasized that although the methods of measurement may be different for the substitution-inert as against the labile ions, the same kind of data can be obtained for the former class as for the latter, and there are moreover a number of powerful methods which as yet are applicable only to substitution-inert ions.

A comparison of interchange rates analogous to those shown in Table I-3 can be made for the substitution-inert cation $Co(NH_3)_5H_2O^{3+}$. For the species $Co(NH_3)_5H_2O^{3+} \cdot H_2O$,

$Co(NH_3)_5H_2O^{3+} \cdot SO_4^{2-}$, and $Co(NH_3)_5H_2O^{3+} \cdot H_2PO_4^-$ the specific rates at 25° are $0.6 \times 10^{-5}$, $0.2 \times 10^{-5}$, and $0.7 \times 10^{-6}$ $sec^{-1}$ [23]. These rates, just as is the case for the appropriate comparisons in Table I-3, can be considered to be remarkably alike. Differences are noted, but these are hardly outside the range of those noted in Table I-3. Here, too, the agreement in rates tempts one to conclude that the bond to the incoming group is made only after the bond to the common leaving group is severed. But for these ions it is known with certainty that the reactions do not take place by limiting $S_N1$ mechanisms. The background for this conclusion is worth going into because it bears directly on the cautionary remarks made in the preceding paragraph.

Haim and Taube [24] studied the product ratio

$$Co(NH_3)_5X^{2+}/Co(NH_3)_5H_2O^{3+}$$

when the reaction

$$Co(NH_3)_5N_3^{2+} + HONO + H^+ \rightarrow [Co(NH_3)_5N_2^{3+}] + N_2O + H_2O$$

takes place in the presence of the ligand $X^-$. The rationale of the experiments was that $N_2$ can be expected to be a very good leaving group and therefore should be as ready as any to leave the coordination sphere of the metal ion without assistance from the entering group. If an $S_N1$ process is assumed for the decomposition of the nitrogen complex, the product ratios lead to values of relative rates for the reactions

$$Co(NH_3)_5^{3+} + X^- \xrightarrow{k_x} (Co(NH_3)_5X^{2+} \qquad (I-8)$$

$$Co(NH_3)_5^{3+} + H_2O \xrightarrow{k_w} Co(NH_3)_5X^{2+} \qquad (I-9)$$

The ratios $k_x/k_w$ for $X^- = Cl^-$, $Br^-$, $NO_3^-$, and $NCS^-$ were found to be 0.32, 0.37, 0.44, and 0.4. The reactivity ratios $k_x/k_w$ can also be calculated from data on the spontaneous anation of

$Co(NH_3)_5H_2O^{3+}$, if it is assumed that an $S_N1$ mechanism operates. The ratios calculated from the data obtained for the first three ions are 0.35, 0.43, and 0.38. Within the accuracy of the measurements, which was not high for the nitrosation system, the reactivity ratios for the two different ways of generating the intermediate could be considered to be the same, and the approximate agreement, moreover, was observed to extend also to $H_2PO_4^-$ and $SO_4^{2-}$ as entering groups. On this basis, it was concluded that the reactions involve the common intermediate $Co(NH_3)_5^{3+}$.

This conclusion, however, did not survive the first test of its capacity to predict product compositions in an ion–ion replacement reaction. Pearson and Moore [25] measured the yield of $Co(NH_3)_5NCS^{2+}$ vs. $Co(NH_3)_5H_2O^{3+}$ when $Co(NH_3)_5NO_3^{2+}$ reacts in aqueous $NCS^-$, and found it to fall far short (by a factor of at least 7) of that expected on the basis of an $S_N1$ mechanism using the reactivity ratios of Haim and Taube. The conclusion is inescapable that despite the agreement noted by Haim and Taube between the nitrosation and spontaneous anation systems, the reactions do not take place by $S_N1$ mechanisms in both systems, though the possibility remains open that this kind of mechanism does operate in one of them. The measurements by Haim and Taube have since been refined by Buckingham *et al.* [26], who report 0.25 $\pm$ 0.02, 0.26 $\pm$ 0.02, and 0.49 $\pm$ 0.02 for $Cl^-$, $Br^-$, and $NO_3^-$, respectively, in the nitrosation of the azido complex. These ratios, though still near those for the spontaneous anation reaction, are now recognizably different, and there is now no reason to assume identical mechanisms. For the reaction of $Co(NH_3)_5O_2CNH_2^{2+}$ with $HONO_2$, in which $CO_2$ is presumably formed as leaving group, the competition ratios for $Cl^-$, $Br^-$, and $NO_3^-$ are observed to be 0.24 $\pm$ 0.02, 0.24 $\pm$ 0.02, and 0.49 $\pm$ 0.02. The close agreement of the precise reactivity ratios for $N_2$ and $CO_2$ as leaving groups does strongly suggest a common intermediate for these two systems, hence $S_N1$ mechanisms for these reactions, and we can conclude, therefore, that the spontaneous anations do not take place by clean $S_N1$ processes. This conclusion then applies to the interchange rates mentioned above, and the general conclusion that caution must be exercised in inferences about mechanisms on approximate rate agreements applies also to the labile ions.

Quite apart from their usefulness in deciding on mechanisms of reactions, the rate comparisons are important in themselves and can often be used in an empirical way to estimate the rate of a substitution reaction. One such application was made by Malin and Swinehart [27]. The rate of replacement of $H_2O$ from $V(H_2O)_6^{2+}$ by $NCS^-$ and of $H_2O$ from $Ni(H_2O)_6^{2+}$ by both $H_2O$ and $NCS^-$ is known and, on assuming that the rates for the reactions

$$M(H_2O)_6^{2+} + H_2O^* \rightarrow M(H_2O)_5H_2O^{* \, 2+} + H_2O \qquad (I\text{-}10)$$

$$M(H_2O)_6^{2+} + NCS^- \rightarrow M(H_2O)_5NCS^+ + H_2O \qquad (I\text{-}11)$$

would bear the same ratio for M = V as for M = Ni, they calculated the water-exchange rate constant for $V(H_2O)_6^{2+}$ as 122 $sec^{-1}$, in excellent agreement with the observed value.

Comparisons of the kind illustrated above are likely to be more successful for dipositive ions than for tripositive ions, because the former, having a lower ionic potential, can be expected to discriminate less between different nucleophiles than do the latter. A few comparisons are made for tripositive ions in the following.

The specific rates for $H_2O$, $NCS^-$ [28], and $Cl^-$ [29] replacing $H_2O$ in $Cr(H_2O)_6^{3+}$ are $2.8 \times 10^{-6}$ $sec^{-1}$, $1.8 \times 10^{-6}$ $M^{-1}\,sec^{-1}$ (at $\mu = 1.00$), and $3 \times 10^{-8}$ $M^{-1}\,sec^{-1}$ (at an anion concentration of $4.4\,M$). For $NCS^-$ [30] and $Cl^-$ [31] replacing $H_2O$ in $Fe(H_2O)_6^{3+}$, these specific rates are 127 ($\mu = 0.40$) and 9 ($\mu = 1.00$).

Assuming that the relative rates of substitution by $NCS^-$ and $H_2O$ on $Fe^{3+}$ and on $Cr^{3+}$ are the same, the rate of water exchange for the former is estimated as $2 \times 10^2$; using $Cl^-$ for the same kind of comparison, it is estimated as $9 \times 10^2$. The differences are not large, and the values are consistent with the results of the most recent NMR measurements. It is in fact remarkable that the rate comparisons hold up as well as they do over a range of $10^8$ in the lability of the central metal ion.

An upper limit of $2M^{-1} \sec^{-1}$ has been set [32] on the rate of replacement of $H_2O$ by $Cl^-$ in $Co(H_2O)_6^{3+}$, and the rate comparison with $Fe(H_2O)_6^{3+}$ is the basis for the upper limit given in Table I-1 for the rate of water exchange for the former ion. Unfortunately, a lower limit on the rate of exchange cannot as yet be placed; in the only experiments [33] done to measure this rate directly, electron-transfer catalysis [34] of water exchange was so prominent as to vitiate the purpose of the measurement.

The rate of replacement of $H_2O$ in $V(H_2O)_6^{3+}$ by $NCS^-$ has been measured [35a] as $114 \pm 10$ at $25°$ and $\mu = 1.00$, and this value has been used in estimating the rate of water exchange for $V(H_2O)_6^{3+}$. It needs to be pointed out that the estimate in this case is especially hazardous because the kinetic evidence points to the conclusion that there is a considerable amount of bond making in the activated complex for substitution.

Ammine complexes have played an important part in developing some of the basic mechanistic principles in the field of electron transfer in solution. The reason for their special place is that many ammine complexes used as reagents retain their integrity for long periods of time in aqueous solution. Furthermore, of the common ligands, $NH_3$ has the lowest capacity to facilitate electron transfer, and is therefore a useful blocking group in mixed complexes where the role of the heteroligand in the electron-transfer process is being defined. In general, release of $NH_3$ from a metal ion is considerably slower than release of water. For example, aquation of $NiNH_3^{2+}$ is governed by a rate constant [35b] of $5 \sec^{-1}$, while for water exchange the corresponding specific rate is $3 \times 10^4 \sec^{-1}$. The specific rate for water exchange on $Cr(H_2O)_6^{3+}$ is $3 \times 10^{-6}$, while for loss of $NH_3$ [36]

(after correction by the statistical factor of 6 so as to conform to the water-exchange rate) it is $2.7 \times 10^{-8}$ sec$^{-1}$. The rate difference when the same comparison is made for $Co(H_2O)_6^{3+}$ and $Co(NH_3)_6^{3+}$ is likely very much greater than for $Cr^{3+}$. In view of the facility with which substitution in $Co(H_2O)_6^{3+}$ takes place, the rate of water exchange is likely to be only a little less than the upper limit of 100 sec$^{-1}$ set by the work on the rate of formation of $CoCl^{2+}$. A solution of $Co(NH_3)_6^{3+}$ in water, if acidified and shielded from light, appears to be indefinitely stable, and an upper limit on the rate of replacement of $NH_3$ by water of $10^{-8}$ sec$^{-1}$ is indicated. The quantitative disparity in the behavior of the $Cr^{3+}$ and $Co^{3+}$ systems will be referred to again.

A reason for the difference in the lability for bound $NH_3$ compared to $H_2O$ may be that $H_2O$, having an electron pair exposed, is sensitive to the influence of electrophilic external groups. The nitrogen of the coordinated ammonia is shielded from such effects, and more stretching of the $M-NH_3$ bond is required before surrounding groups can interact with the leaving ligand.

## V. COMMENTS ON RATES SUMMARIZED IN TABLE I-1

The influence of electronic structure on the substitution rates of metal ions has been strongly emphasized [37, 38], perhaps to such an extent that other factors affecting the rates are often neglected. An important comparison which shows the influence of the charge difference is that between $V(H_2O)_6^{2+}$ and $Cr(H_2O)_6^{3+}$. Each of these ions has the $\pi d^3$ electron structure, but the lability of $V(H_2O)_6^{2+}$ is about $10^8$ times as great as that of $Cr(H_2O)_6^{3+}$. However, this ratio is by no means constant for $2+$ compared to $3+$ ions of the same electronic structure. Thus, when $Mn(H_2O)_6^{2+}$ is compared to $Fe(H_2O)_6^{3+}$, the ratio is about $10^5$. The fact that the ratio changes may be disappointing, but should not be too surprising. The rate, even in the absence of covalent bond or crystal-field effects, can be expected to be a sensitive function of the way water molecules fit around a cation; and a small change in radius, when it occurs at a distance critical for a change in coordination number, can have a profound effect on the lability.

The influence of electronic structure on the rate of substitution has been discussed in terms of the valence-bond theory and also in terms of crystal-field theory. As in describing other properties of complex ions, a description that encompasses bond covalency and the influences of ligands on nonbonding electrons is potentially more powerful than either. However, none of the approaches is yet useful quantitatively, but each has a range of usefulness in qualitative discussions. The valence-bond approach has the obvious defect that it takes no account of the nonbonding electrons, except as they exclude certain metal-centered orbitals from participating in forming the activated complex. It should be mentioned in this connection that, contrary to assertions which have been made, the valence-bond approach does not predict an $S_N2$ mechanism for an octahedral complex such as $V^{3+}$, which has a $\pi d^2$ electron structure, and thus an unoccupied $d$ orbital, though in view of the evidence adduced by Baker et al. [35], it would be tempting to make this claim for it. An $S_N2$ complex would be easier to form for a $\pi d^2$ than for a $\pi d^3$ electronic state; but it is also true that if the coordination number of the activated complex is less than that of the ground state, there is a compensating influence for the $\pi d^2$ case, in that a $d$ orbital of lower quantum number can be used in the rehybridization. The crystal-field approach in turn has the obvious defect that it does not explain the differences in lability among $Mg^{2+}$, $Al^{3+}$, $Si^{4+}$, etc. The purely electrostatic contribution to the activation energy can be very large, and these are neglected except as they affect the energy states of nonbonding electrons in the valence shell. The crystal-field approach is, however, the more useful for transition metals of low charge, because it at least offers an opportunity for understanding the differences in lability between $Fe^{2+}$, $Co^{2+}$, and $Ni^{2+}$, even though it is incapable of exploiting the opportunity fully.

An extremely important factor affecting labilities is the structure of the complex ion in the ground state. All of the deliberations mentioned have been based on the tacit assumption that the ions compared have the same radius and shape. The comparisons founder to some extent on the point that, even for dipositive ions of the first row, the radii change. Even more dramatic is the effect of change in shape (which itself may be largely a reflection of change in electronic structure). Thus $Cu(H_2O)_4(H_2O')_2^{2+}$ and $Cr(H_2O)_4(H_2O')_2^{2+}$ among the dipositive ions of the first row are remarkably labile. For this to be understood, we must take account of the fact that both ions depart strongly from regular octahedral shapes; as has been mentioned, it is possible that substitution at the axial positions is very facile, and the overall substitution rate is determined by the rate of pseudorotation within the complexes. These rates need bear no simple relation to electronic structures along the lines of the claims which have been made for rates of substitution.

The dramatic rate differences observed can be rationalized qualitatively in simple terms. For example, the enormous increase in lability between $Cr(H_2O)_6^{3+}$ and $Cr(H_2O)_6^{2+}$, going beyond what seems reasonable for only a change in charge, can be ascribed to the fact that an antibonding electron is added when Cr(III) is reduced; the same factor accounts for a large part of the rate difference between Co(III) and Co$^{2+}$. The relatively small rate difference between $V(H_2O)_6^{2+}$ and $V(H_2O)_6^{3+}$ is attributable to the fact that the change in electronic structure from V(II) to V(III) ($\pi d^3$ to $\pi d^2$) favors substitution on V(III), but this is largely offset by the difference in charge which favors substitution on V(II).

## VI. PROBLEMS

The progress which has been made in the last two decades in ion solvation is remarkable, but the field is so extensive and diverse that much remains to be done. The problems which will be mentioned do not involve mere refinements of work that has already provided solutions in a first approximation but, in fact, call for first-order solutions.

A field of research in ion solvation which is extremely active deals with non-aqueous solvents and with solvent mixtures, particularly those having water as a component. For many of these systems, the liquid range extends to temperatures far below $0°C$, and the power of the methods which depend on kinetic effects is thereby increased. Not all non-aqueous solvents show the strong background absorption in the infrared which is shown by water, and this technique comes into its own at least in studies of the structure of the first coordination sphere. Progress in the field of solvation in non-aqueous solvents is being made rapidly, and the ground is being prepared for a fundamental attack on the mechanism of electron transfer in non-aqueous or partially aqueous solvents.

An extremely important subject in ion solvation deals with species in which water is partially displaced from the coordination sphere of a metal ion by another ligand. The interest here is twofold: to learn the number and disposition of the residual water molecules (this problem is particularly acute for poly-dentate ligands), and to learn the effect of the other ligands on the lability of the coordinated water molecules. Both of these problems are enormous in extent, and both are basic to under-standing the mechanisms of redox processes.

Even for the purely aqueous systems, challenging problems remain. A large number of diamagnetic aquo cations, among them $Tl^+(aq)$, $Tl^{3+}(aq)$, $Hg^{2+}(aq)$, $Hg_2^{2+}$, $Ag^{2+}(aq)$, $Ag^+(aq)$, and $Sn^{2+}(aq)$, have not been characterized, even though they have figured prominently in studies of oxidation–reduction reactions. The solvation of rare earth ions and of the actinides is not at all well understood, though progress is in prospect for the former class from a study of the spectra. The rare earth ions, $Tl^{3+}$, $Hg^{2+}$, and $Sn^{2+}$ will probably yield to one of the techniques which has been described, but for $Tl^+$, $Ag^+$, and probably $Hg_2^{2+}$

the problems may not be amenable to solutions of the kind which apply to many cations. When the electric field intensity of the cation is small enough, the energy difference between different compositions and/or structures of the aquo cations may be so small that several forms must be taken into consideration in describing the ground states of the system.

The general problem in understanding the lability of different aquo ions has been referred to. One ion, Co(III), has figured so prominently in studies of mechanisms of electron transfer that its idiosyncrasy with respect to substitution properties deserves special mention. I draw attention here to the enormous difference in lability of the species $Co(H_2O)_6^{3+}$ and $Co(NH_3)_6^{3+}$. The difference between $Cr(H_2O)_6^{3+}$ and $Cr(NH_3)_6^{3+}$, or between $Ni(H_2O)_5NH_3^{2+}$ and $Ni(H_2O)_6^{2+}$, is perhaps more normal, and on Cr(III) or Ni(II) the difference in rate of release of $H_2O$ compared to $NH_3$ is much less than for Co(III). This abnormally high lability of $Co(H_2O)_6^{3+}$ compared to $Co(NH_3)_6^{3+}$ was noted quite some time ago [37], though perhaps without a substantial experimental basis. It seems reasonable that the interpretation offered then is still applicable to the situation: the splitting between the low-spin ground state and the high-spin labile electronic state is small enough for the aquo ion so that substitution can take place through the high-spin labile state.

## REFERENCES

1. Values are taken from W. M. Latimer, "Oxidation Potentials," 2nd Ed., Prentice-Hall, Englewood Cliffs, New Jersey, 1952, except as otherwise noted, but are expressed in conformity with the Stockholm convention.
2. C. W. Merideth, Ph.D. Thesis, University of California, 1965.
3a. J. P. Hunt and H. Taube, *J. Chem. Phys.* 19, 602 (1951).
3b. J. P. Hunt and H. Taube, *J. Chem. Phys.* 18, 757 (1950).
4. J. P. Hunt and R. A. Plane, *J. Am. Chem. Soc.* 76, 5960 (1954).
5. M. V. Olson, Y. Kanizawa, and H. Taube, *J. Chem. Phys.*, in press (1969).
6. Estimated from rate comparisons, *vide infra*.
7. T. J. Meyer and H. Taube, *Inorg. Chem.* 7, 2369 (1968).

8. J. A. Stritar, Ph.D. Thesis, Stanford University, Stanford, California, 1967.
9. Based on rate comparison. The rates of aquation of some pentaammine-ruthenium(III) complexes have been measured. J. A. Broomhead, F. Basolo, and R. G. Pearson, *Inorg. Chem.* 3, 826 (1964).
10. T. J. Swift and R. E. Connick, *J. Chem. Phys.* 37, 307 (1962).
11. A. C. Rutenberg and H. Taube, *J. Chem. Phys.* 20, 825 (1952).
12. H. R. Hunt and H. Taube, *J. Am. Chem. Soc.* 20, 2692 (1958).
13. The equilibrium value for the half-reaction $H_2O + Co^{2+} + 5NH_4^+ = Co(NH_3)_5OH_2^{3+} + 5H^+ + e^-$ is about 3.0 V. The driving force for the reduction of the Co(III) complex derives in large part from the affinity of $NH_3$ for $H^+$. However, the kinetic data show that $H^+$ does not engage $NH_3$ in the activated complex for the reduction of Co(III). The value of $E^0$ entered is a "kinetic" value as estimated by A. Haim and H. Taube, *J. Am. Chem. Soc.* 85, 1 (1963).
14. R. E. Connick, private communication.
15. The notation Ce(IV) rather than $Ce^{4+}$ is chosen because the ion hydrolyzes quite strongly in acid, and the constitution of the ion is sensitive to acidity.
16. O. Schmitz-Dumont, H. Brokopf, and K. Burkhardt, *Z. Anorg. Allgem. Chem.* 295, 7 (1958).
17. H. Baldwin and H. Taube, *J. Chem. Phys.* 33, 206 (1960).
18. R. E. Connick and D. N. Fiat, *J. Chem. Phys.* 39, 1349 (1963).
19. R. E. Connick and D. N. Fiat, *J. Chem. Phys.* 44, 4103 (1966).
20. J. A. Jackson, J. F. Lemons, and H. Taube, *J. Chem. Phys.* 32, 553 (1960).
21. This has largely resulted from advances which Eigen and co-workers have made in developing methods for the measurements of rates of rapid reactions in solution. Many of these (and other results as well) are compiled in M. Eigen and R. G. Wilkins, Kinetics and Mechanisms of Formation of Metal Complexes, *Advan. Chem.* 49, 58 (1965).
22. Data are taken from [21], except for $V(H_2O)_6^{2+} \cdot H_2O$, which is from [5] and for $V(H_2O)_6^{2+} \cdot SO_4^{2-}$, which is mentioned in M. Eigen, *Bunsenges. Phys. Chem.* 67, 753 (1963), but without detail on the source of the data.
23. W. Schmidt and H. Taube, *Inorg. Chem.* 2, 698 (1963).
24. A. Haim and H. Taube, *Inorg. Chem.* 2, 1199 (1963).
25. R. G. Pearson and J. W. Moore, *Inorg. Chem.* 3, 1334 (1964).
26. D. A. Buckingham, I. I. Olsen, A. M. Sargeson, and H. Satrapa, *Inorg. Chem.* 6, 1027 (1967).
27. J. M. Malin and H. H. Swinehart, *Inorg. Chem.* 7, 250 (1968).
28. C. Postmus and E. L. King, *J. Phys. Chem.* 59, 1216 (1955).
29. T. W. Swaddle and E. L. King, *Inorg. Chem.* 4, 532 (1965); H. S. Gates and E. L. King, *J. Am. Chem. Soc.* 80, 5011 (1958).
30. J. F. Below, Jr., R. E. Connick, and C. P. Coppel, *J. Am. Chem. Soc.* 80, 2961 (1958).

31. R. E. Connick and C. P. Coppel, *J. Am. Chem. Soc.* **81**, 6389 (1959).
32. T. J. Conocchioli, G. H. Nancollas, and N. Sutin, *Inorg. Chem.* **5**, 1 (1966).
33. H. L. Friedman, H. Taube, and J. P. Hunt, *J. Chem. Phys.* **18**, 759 (1950).
34. N. A. Bonner and J. P. Hunt, *J. Am. Chem. Soc.* **82**, 3826 (1960).
35a. B. R. Baker, N. Sutin, and T. J. Welsh, *Inorg. Chem.* **6**, 1948 (1967).
35b. G. A. Melson and R. G. Wilkins, *J. Chem. Soc.* 4208 (1962).
36. J. Bjerrum and C. S. Lamm, *Acta Chem. Scand.* **9**, 216 (1955); E. Jorgenson and J. Bjerrum, *ibid.* **12**, 1047 (1958.
37. H. Taube, *Chem. Rev.* **50**, 69 (1952).
38. F. Basolo and R. G. Pearson, "Mechanisms of Inorganic Reactions," 2nd Ed. Wiley, New York, 1967.

# II

## DESCRIPTION OF THE ACTIVATED
## COMPLEXES FOR ELECTRON TRANSFER

### I. INTRODUCTION

The modern era of research on the mechanism of electron transfer in solution began when artificially produced radioactive elements became available. The new nuclei made it possible to measure the rates of many exchange reactions, and these measurements attracted attention not only because of their novelty, but also because of the bearing the results had on understanding orthodox chemical reactions. The rate of exchange between $Fe^{2+}$ and $Fe^{3+}$ in solution came in for early attention [1], and studies on this system have played a key role in the development of the subject of electron transfer in solution [2].

Ideas on the possibility, on the one hand, that electron transfer might occur through intact coordination shells, or, on the other, that it might be mediated by some atom shared between the coordination shells were current quite early [1] among those who concerned themselves with these problems, but they were not clearly formulated in the literature until the proceedings were published of a symposium held in 1951 on electron-transfer processes. The distinction between inner-sphere [3] and outer-sphere [4] activated complexes was clearly made in the course of the discussion of some of the experimental results presented at the symposium, although the names used here to describe the two types of mechanism were not introduced until later. The distinction is fundamentally between reactions in which electron transfer takes place from one primary bond system to another (outer-sphere mechanism), and those in which electron transfer takes place within a single primary bond system (inner-sphere mechanism). The classification will be used in this chapter, which is devoted to describing the experimental basis for the conclusions about mechanism in some representative systems, and some of the properties of the two kinds of activated complexes. As will appear from the discussion, some refinement or elaboration of the primitive classification is already required by the experimental results, and this process will undoubtedly continue as we learn more about these reactions and understand their mechanisms better.

## II. The Outer-Sphere Activated Complex

### A. General

There are many complexes which undergo electron transfer more rapidly than they undergo substitution. When a redox reaction occurs between two such agents, electron transfer must take place through the intact coordination spheres of the reaction partners. This conclusion is, of course, based on the proviso that neither agent labilizes the other for substitution, and it is difficult to see how a labilization effect could occur except by electron transfer. Even when one of the reaction partners is substitution-labile, if the other partner does not offer a suitable site to engage the metal ion of the labile partner, reaction by an outer-sphere activated complex will be favored. This situation is encountered in the reduction of $Co(NH_3)_6^{3+}$ [5], or of $Co(NH_3)_5py^{3+}$ [6, 7] by $Cr^{2+}(aq)$; despite the propensity which

$Cr^{2+}$ has to react by inner-sphere mechanisms, and the lability of its aquo complex, the oxidizing agents mentioned are reduced by outer-sphere mechanisms. Coordinated $NH_3$ has no unshared electrons to provide a means of attachment to $Cr^{2+}$, and although the point is less obvious for pyridine, the same situation seems to apply here as well. In basic solution, when $NH_3$ is converted to $NH_2^-$, an inner-sphere mechanism for the reduction of ammine complexes may become important.

A large number of reactions taking place by outer-sphere mechanisms have by now been studied in some detail [8, 9], both self-exchange [10] and cross reactions [10], and for all at constant ionic composition the rates are simply first-order in each reactant, or perhaps it is more correct to say that we shall restrict our attention to the large majority which show this simple kind of kinetic behavior. In all cases thus far studied involving the reaction of octahedral complexes in which only $\pi d$ orbitals are populated, the rates are quite rapid, with $k$ of the order of 1 $M^{-1}$ sec$^{-1}$ or greater. Some remarkable salt effects have been observed for outer-sphere reactions. They are illustrated in Table II-1 for the $Fe(CN)_6^{4-,3-}$ self-exchange reaction [11], a system of extreme charge type.

TABLE II-1

SALT EFFECTS IN THE $Fe(CN)_6^{4-,3-}$ SELF-EXCHANGE REACTION[a]

| Cation | $(CH_3)_4N^+$ | $(n{-}C_3H_7)_4N^+$ | $(n{-}C_5H_{11})_4N$ | $Co(C_5H_5)_2^+$ | $K^+$ |
|--------|------|------|------|------|------|
| $k$ | 1260 | 41 | 16 | 1050 | 230 |
| $E$ | 0.3 | 5.2 | 7.8 | —1 | 6.0 |

[a] Reactions at 0.1°, cations at 0.01$M$.

Only a brief and general account is being made of what is known about outer-sphere reactions, but it cannot be concluded without mention of the important contribution which Marcus

[*12*] has made in formulating a rule for correlating the rates of self-exchange and cross reactions. Exceptions [*13, 14*] to the correlation have been noted, but they are not so numerous as to disqualify the correlation as a touchstone of normal behavior. The significant exceptions involve the Co(II)—Co(III) couple, and this couple is unusual in involving a drastic electron reorganization on change in oxidation state.

## B. *Problems*

Though rate laws are known, activation parameters have been measured, the effects of salts on rates have been determined, and a useful rate correlation has been introduced, fundamental questions about the mechanism of electron transfer for outer-sphere reactions remain to be answered. Important among these is the question of the distance of approach which is optimum for electron transfer. Are the ions in contact on electron transfer, or are there intervening solvent molecules or counter ions in the configuration which is most favorable for electron transfer? If there are differences in configuration for different systems, how is the configuration affected by changes in the properties of the complexes, of the solvent, or of the counter ions? It seems possible, for example, that when water is the solvent, the ions do actually come into contact, but also it seems possible that, in a solvent such as pyridine, electrons can transfer through the medium. Because the associations distinguishing one configuration of an activated complex from another are very labile ones, it will be difficult to frame simple and direct experiments which can provide unambiguous answers to the questions which have been posed. Progress in understanding them will, in all likelihood, depend on an interplay of rate comparisons for a large variety of systems with theory along the lines pursued by Marcus and others.

Another and related group of questions deals with salt effects. Do the salt effects illustrated in Table II-1 mean that the counter ions can act as charge carriers? An interpretation of this kind is invited by the fact that the influence of the counter ion can in some cases be expressed by a rate law [15, 16] (this implies that the activated complex has a definite molecularity with respect to the counter ion in question) and by the proof [17] that alkali metal ions do serve as charge carriers in some reactions of organic radical ions in solvents of low dielectric constant. But it must be recognized that there is a vast difference quantitatively between $Cs^+$ mediating in electron transfer between $MnO_4^-$ and $MnO_4^{2-}$, and $Na^+$ mediating between

In the latter system, the radical ion is almost as good a reducing agent as is the alkali metal; that is, the states $Na^+ \cdot Arom^-$ and $Na \cdot Arom$ do not differ much in energy, but there is an enormous energy difference between the states $Cs^+ \cdot MnO_4^{2-}$ and $Cs \cdot MnO_4^-$, and this energy difference may well disqualify $Cs^+$ as operating in the way that $Na^+$ does for the organic system. It is quite possible that the remarkable salt effects which have been observed in some electron-transfer reactions have less to do with the fact that electron transfer is involved than with the fact that electron-transfer reactions often feature extreme charge types, so that activated complexes of unusually high charge are encountered. Substitution reactions require very close contact of the reaction partners, and they tend to become very slow when the partners have charges of the same sign.

The influence which changes in the ligands exert on the rates of outer-sphere reactions is by no means fully understood. Some of the problems raised in this connection are illustrated in Table II-2, and will be outlined and considered.

TABLE II-2

RATES OF ELECTRON TRANSFER FOR OUTER-SPHERE REACTIONS
AS A FUNCTION OF LIGAND STRUCTURE

| System | Oxidant | Reductant | Total anion concentration | $k$ $(M^{-1}\,sec^{-1})$ | Reference |
|--------|---------|-----------|---------------------------|--------------------------|-----------|
| 1 | $Ru(o\text{-phen})_3^{3+}$ | $Ru(o\text{-phen})_3^{2+}$ | 0.20 | $>10^7$ [a] | [18] |
| 2 | $Ru(ND_3)_6^{3+}$ | $Ru(NH_3)_6^{2+}$ | 0.01 | $8 \times 10^2$ | [19] |
| 3 | $Co(NH_3)_6^{3+}$ | $Co(NH_3)_6^{2+}$ | — | $<3 \times 10^{-12}$ | [20] |
| 4[b] | $Co(o\text{-phen})_3^{3+}$ | $Co(o\text{-phen})_3^{2+}$ | 0.10 | $5.0$ [c] | [21] |
| 5 | $Co(NH_3)_6^{3+}$ | $Cr(bipy)_3^{2+}$ | 0.20 | $2.5 \times 10^2$ | [22] |
| 6[d] | $Co(ND_3)_6^{3+}$ | $Cr(bipy)_3^{2+}$ | 0.20 | $1.9 \times 10^2$ | [22] |
| 7[e] | $Co(NH_3)_5OH_2^{3+}$ | $Cr(bipy)_3^{2+}$ | 0.05 | $2.1 \times 10^3$ | [22] |
| 8[e] | $Co(NH_3)_5OH^{2+}$ | $Cr(bipy)_3^{2+}$ | 0.05 | $1.2 \times 10^3$ | [22] |
| 9[e,f] | $Co(ND_3)_5OD_2^{3+}$ [d] | $Cr(bipy)_3^{2+}$ | 0.05 | $8 \times 10^2$ | [22] |
| 10[e,f] | $Co(NH_3)_5OD_2^{3+}$ [d] | $Cr(bipy)_3^{2+}$ | 0.05 | $8 \times 10^2$ | [22] |

[a] The measurement was actually made for $Fe(o\text{-phen})_3^{2+,3+}$ but it can be assumed rather confidently that the rate for the corresponding Ru complexes will be about as great.

[b] The rates of exchange for the $o$-phen and bipy complexes have been shown to be almost identical [21].

[c] At 0°.

[d] In $D_2O$; in a separate experiment the isotopic change in the solvent was shown to produce an insignificant change in rate.

[e] At 4°.

[f] In $D_2O$.

An immediate question which is raised when the first two systems are compared is the cause of the great rate difference indicated. (Comparisons made elsewhere in Table II-2 show

that the difference can hardly be ascribed to the fact that D
replaces H in one of the reaction partners in the rutheniumam-
mine system.) Is the greater rate for system 1 a result of the
greater bulk of the ligands in this case (this will have the effect
of diminishing the interaction of the ions with the solvent and
thus decreasing the energy which needs to be expended prior to
electron transfer in reorganizing the solvent) or to the fact
that the change in the Ru—N distance is less for system 1 than
for system 2 (this cannot be stated as fact, but it seems very
likely that when the measurements are made, there will prove to
be at least a slight difference in the direction indicated) or is it
a result of the greater electron delocalization for system 1? The
questions posed cannot yet be given a satisfactory answer. All
three factors undoubtedly contribute, and a satisfactory answer
would require a quantitative assessment of the contribution made
by each. An attempt will be made at a qualitative assessment.

Self-exchange in $Ru(en)_3^{2+,3+}$ takes place at a slightly lower
rate than for $Ru(NH_3)_6^{2+,3+}$ [19], and thus it is unlikely that the
greater bulk of the ligand in system 1 accounts for the greater
rate of self-exchange observed for the system. It is also unlikely
that the Ru—N distance is an important consideration. The
change in Co—N distance when $Co(NH_3)_6^{3+}$ is reduced to
$Co(NH_3)_6^{2+}$ is only [23] 0.15 Å, so that the dimensions come close
to overlapping, or indeed do overlap in the course of the zero-
point vibrations. The change in dimensions for the cobaltammine
system as the oxidation state is changed is expected to be unusually large because Co(II) has two antibonding $d$ electrons, while
Co(III) has none, and a much smaller change in dimension for
the $Ru(NH_3)_6^{2+,3+}$ system is expected because here neither state
has antibonding $d$ electrons. If any of the three factors considered
is the major cause of the rate difference under discussion, it is
likely the third one, but the point needs to be investigated
further before the conclusion is accepted.

The comparisons for systems 3 and 4 show that the change
from $o$-phenanthroline ($o$-phen) to $NH_3$ as ligand causes a much

greater rate difference in the Co(II)-Co(III) than it does in the Ru(II)-Ru(III) couple. A qualitative reason for this difference can be offered based on the fact that the cobalt couple has a complication not featured by the other. Simple electron transfer between $Co(NH_3)_6^{2+}$ and $Co(NH_3)_6^{3+}$ would leave both reaction partners in excited electronic states ($Co^{3+}$ would be left as $t_{2g}^4 e_g^2$ rather than as $t_{2g}$, while $Co^{2+}$ would be left as $t_{2g}^6 e_g^1$ rather than as $t_{2g}^5 e_g^2$). The energy requirement is less severe if one of the reaction partners undergoes electronic excitation prior to reaction. For $Co(o\text{-phen})_3^{2+}$, though the ground state is high-spin, the low-spin state probably is only slightly higher in energy, and the added energy requirement is not very costly in this case. For the cobaltammine system, whether $Co^{2+}$ or $Co^{3+}$ is excited before reaction, the expenditure in energy is much greater, and the rate of the self-exchange reaction is correspondingly less.

Another problem raised by the data in Table II-2 is the smaller rate of reduction for $Co(NH_3)_6^{3+}$ compared to $Co(NH_3)_5OH_2^{3+}$. A part of the difference can, in terms of the Marcus correlation [12], be ascribed to the difference in driving force because the pentammine complex is probably a stronger oxidant than the hexammine. However, the isotopic experiments suggest that a more specific factor may also be involved. The entries for systems 5 and 6 compared to those for 9 and 10 show that the reaction rate is quite insensitive to replacement of H by D on $NH_3$, but very sensitive to the corresponding replacement on $H_2O$. To understand the sensitivity of the rate to isotopic substitution on water, this ligand water must be able to assume a special role not open to $NH_3$, and this in turn implicates the unshared electron pair in the coordinated molecule of water, even though an outer-sphere mechanism operates.

The effect of ligands on the rate of electron transfer is also illustrated by the comparison of $Co(NH_3)_5OH_2^{3+}$ and $Co(NH_3)_5OH^{2+}$ as oxidants, the former being the more reactive to $Cr(bipy)_3^{2+}$. Now a molecule of water in the coordination sphere

of Co(III) is about $10^4$ times more acidic than it is on Co(II), so
that the hydroxo complex is expected to be a weaker oxidant than
the aquo complex. In terms again of the Marcus correlation
[12], this factor alone can account for a rate decrease of about 100
for the hydroxo complex, and it must be supposed that unless
still another factor is operative, the change in driving force more
than compensates for the change in electrostatic repulsion.

Little systematic work has been done on the effect of changing
the solvent. The interpretation of the results is likely to be difficult for the common inorganic systems involving ions of high
charge, because the effects of counter ions become very great in
solvents of low dielectric constant. Nevertheless, a systematic
study of the effect of change in solvent seems worthwhile,
particularly if it extends to solvents such as pyridine, the molecules of which have relatively low-lying unoccupied orbitals.

## III. The Inner-Sphere Activated Complex

### A. *Introduction*

The difficulties in defining the geometry of an outer-sphere
activated complex have been ascribed to the great lability of the
interactions which distinguish one configuration from another—
as, for example, a configuration in which a molecule of water is
interposed between $Ru(NH_3)_6^{2+}$ and $Ru(NH_3)_6^{3+}$ and another in
which two water molecules are interposed. Primary bonds are
not disrupted in forming the activated complexes from the
reactants, nor in proceeding from one possible configuration of
the activated complex to another. When primary bonds are
affected in the course of the reaction, as is the case when an inner-
sphere mechanism operates, it can be expected that, for some
systems at least, the effects will persist long enough so that they can
be observed in the products. The success of the demonstration
[24, 25] that electron transfer between metal ions can take place

by an inner-sphere mechanism depended on choosing a system in which the opportunity mentioned could easily be exploited. The reaction studied was the reduction of $Co(NH_3)_5Cl^{2+}$ by $Cr^{2+}$. The system was chosen with the knowledge that the Co(III) complex retains its integrity in solution for a long period of time, that $Cr^{2+}$ is labile to substitution, that $Co^{2+}$ is labile to substitution, that $CrCl^{2+}$ is slow to aquate, and with the hope that $Cr^{2+}$ would reduce Co(III) rapidly. This hope was realized (it has been shown [26] that the specific rate of the reduction reaction exceeds $10^5 \, M^{-1} \, sec^{-1}$ at 25°) and $CrCl^{2+}$ was found to be the main Cr(III) product of the reaction of $Co(NH_3)_5Cl^{2+}$ with $Cr(H_2O)_6^{2+}$. In an experiment with radioactive $Cl^-$ present, it was further shown that very little radioactive chloride is incorporated into the product $CrCl^{2+}$. The experiments prove that, as $Cr^{2+}$ is oxidized by $Co(NH_3)_5Cl^{2+}$, chlorine is transferred directly from Co(III) to Cr, and thus it is proven that in the activated complex $Cl^-$ bridges the Co and Cr atoms, as shown in the structure

$$[(NH_3)_5CoClCr(H_2O)_5]^{4+}$$

In this formulation of the activated complex it has been assumed, though this point has not been proved, that $Co(III)-Cl^-$ occupies a normal coordination position on the chromium. This point will, however, be referred to again and, in the meantime, this assumption will be accepted without further question.

The inner-sphere activated complex has been more completely described in a related reaction:

$$5H_2O + Co(NH_3)_5H_2O^{3+} + Cr(H_2O)_6^{2+} + 5H^+$$

$$= Co(H_2O)_6^{2+} + 5NH_4^+ + Cr(H_2O)_6^{3+} \qquad \text{(II-1)}$$

Kinetic studies with this system [27, 28] show the rate law for the reaction to be

$$-d[Co(NH_3)_5H_2O^{3+}]/dt = [Cr^{2+}][Co(NH_3)_5H_2O^{3+}]\{k_1 + k_2/[H^+]\}$$

where the second term may be expressed in the kinetically equivalent form

$$k_2'[\text{Cr}^{2+}][\text{Co(NH}_3)_5\text{OH}^{2+}].$$

At 25° and $\mu = 0.5$, $k_1$ is 0.6 $M^{-1}$ sec$^{-1}$, and $k_2'$ is $\sim 2 \times 10^6 M^{-1}$ sec$^{-1}$. Chemical identification of intermediate Cr(III) products is no help in establishing the mechanism of reaction (II-1) because $\text{Cr(H}_2\text{O)}_6^{3+}$ is the expected product whether an inner-sphere or outer-sphere mechanism operates. Tracer experiments [28, 29] with $^{18}$O have, however, provided proof that by each of the two paths, oxygen transfer from Co to Cr is essentially quantitative, and proof therefore that $\text{Cr}^{2+}$ attacks the oxygen in the coordination sphere of Co(III).

Isotopic fractionation studies have provided a deeper insight into the nature of the activation process. The fractionation factor d ln $^{16}$O/d ln $^{18}$O for the $k_2$ path in the aquo ion [28], after correction for the fractionation introduced by the operation of the acid dissociation equilibrium [30], is about $1.03_5$. The kinetic isotope effect on the Co−N bonds is very small: d ln $^{14}$N/d ln $^{15}$N is found to be 1.002–1.003, and is not materially different for NH$_3$ in the *cis* and *trans* positions [31]. The fractionation indicates that activation of the oxidizing agent requires considerable stretching of the Co−OH bond, but very little change in the dimension of the molecule otherwise; that is, the bond dislocation is concentrated in the bridging group position. Replacement of H by D in the coordinated NH$_3$ has little effect on the rate of reaction [27], but the solvent change from H$_2$O to D$_2$O has a large effect, both $k_1$ and $k_2$ decreasing by a factor of about 4 [27]. There are three components in the H$_2$O−D$_2$O solvent change: D replaces H in the bridging groups HO$^-$ and H$_2$O, D replaces H in the first coordination sphere of Cr$^{2+}$, and D replaces H in the solvent. The last two factors probably contribute only slightly to the isotopic effect: it should be noted that the rate of reaction between Co(NH$_3$)$_5$Cl$^{2+}$ and Cr$^{2+}$ decreases by only a factor of 1.5 when D$_2$O rather than

$H_2O$ is the solvent [32], so that the seat of the large isotope effect is on the bridging group. The major part (that is a factor of 3 out of the total of 4) of the $H-D$ isotope effect for the $k_2$ path is absorbed by the pre-equilibrium [33]

$$Co(NH_3)_5H_2O^{3+} = Co(NH_3)_5OH^{2+} + H^+$$

but the kinetic $H-D$ isotope effect on the $k_1$ path still appears to be large. The $H-D$ experiments are consistent with the conclusion that the bridging group is much more sensitive to this probe than are the other groups present in the reacting complexes.

B. *Electronic Interpretation of the Activation Process in the Reduction of Cobaltammines by Chromous Ion*

A conclusion which will not be documented in detail here, but which is supported by numerous observations, some of which will be referred to in the next chapter, is this: $Cr^{2+}$ in reaction with amminecobalt (III) complexes [and with Cr(III) complexes] shows a marked preference for reaction by an inner-sphere path rather than by an outer-sphere path. Almost without exception, when a ligand on Co(III) presents a suitably located Lewis base site, group transfer from Co(III) to Cr(II) takes place. An electronic description of the electron-transfer act will be offered which rationalizes the isotopic fractionation data, and though, being qualitative, it cannot be said to have the capacity to predict the generality of the inner-sphere mechanism for $Cr^{2+}-Co(III)$, it does serve to resolve some of the factors contributing to this outcome.

In Fig. II-1 are suggested the electronic structures of the reaction partners (A) before reaction, (B) in the activated complex, and (C) after reaction [the electron now being trapped on Co(III)].

It should be noted that the donor orbital for $Cr^{2+}$ has $\sigma$ symmetry with respect to the ligand–metal axis, and that the acceptor orbital on $Co^{3+}$ has similar symmetry. In Fig. II-1A,

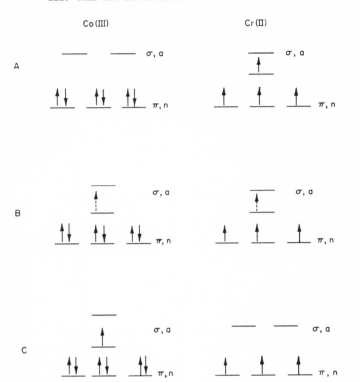

FIG. II-1.   Changes in electronic structure on the reaction of $Cr^{2+}(aq)$ with $Co(NH_3)_5X^{2+}$. A. Before electron transfer. B. In activated complex. C. After electron transfer. The notations $a$ and $n$ refer to antibonding and nonbonding orbitals.

the energy of the orbital containing the electron to be transferred is shown lying lower than the lowest unoccupied orbital for Co(III). For electron transfer to take place, the coordination spheres about Co(III) and Cr(II) must distort so that the electron can transfer without change in energy. The consequences of one such distortion, that of changing the ligand–metal distances along an axis, are shown in Fig. II-1B. Distances on Co are taken to increase, lowering the energy of a $\sigma a$ orbital, and those on Cr are taken to decrease, raising the energy of a $\sigma a$ orbital. When electron transfer takes place by an outer-sphere mechanism, the bond dislocations at the two centers are not coupled, and the improbability that the uncoupled events take place simultaneously when the ions are close enough for electron transfer to take place in all likelihood limits the rate of electron transfer. In an inner-sphere mechanism the dislocations of the two centers are coupled. The movement of the bridging group from Co to Cr— this lowers the energy of a Co orbital and raises that of an orbital centered on Cr—simultaneously satisfies major activation requirements at the two centers, and if the condition of facile electron transfer through a bridging group is satisfied, it provides a way of understanding why $Cr^{2+}$, in reacting with Co(III) complexes, shows such a high selectivity for an inner-sphere mechanism. On the basis of the explanation advanced, this selectivity is a property of the $\sigma$ donor–$\sigma$ acceptor combination, a point which seems to be borne out by the observations.

The description offered for the $Cr^{2+}$-cobaltammine case need, of course, not apply in detail to all others. In fact, an interesting difference with respect to oxygen fractionation in *cis* and *trans* positions has been documented [34] for the reductions of chloro-aquotetramminechromium(III) and of a cobaltpentaammine complex. For the former ion, d ln $^{16}$O/d ln $^{18}$O is 1.017 when water is *trans* to $Cl^-$, and is 1.007 when it is *cis*, and a reasonable explanation for the relatively large value of this factor follows a suggestion made by Orgel: if the incoming electron enters an unhybridized $d_z2$ orbital, the energy can be lowered not only by stretching the bridging ligand—M bond, but also by stretching the $M-NH_3$ bond *trans* to the bridging group. The difference between Co(III) and Cr(III) may rest in the fact that the former ion is the much better oxidizing agent, and total dislocation which is needed to lower the energy of the antibonding orbital can be accommodated in one bond, while for the latter to lower the energy enough, a fluctuation in configuration is needed in which both bonds along the reaction axis are stretched. The further investigation of the N fractionation factors promises to be a very interesting subject in its own right, because it is by no means certain that the results obtained thus far for the reduction of Co(III) are general for cobalt complexes, nor that the *trans* effect noticed for Cr(III) in the reaction of the aquopentaammine complex is general for all Cr(III) complexes.

## C. *Extending the Field of Inner-Sphere Mechanisms*

The inner-sphere activated complex has by now been demonstrated for the reactions of a large number of metal complexes. These will not be catalogued in detail here, but the emphasis in the present discussion will be on the kind of evidence which can be brought to bear on the problem of mechanism. Nevertheless, it should be mentioned that the catalogue of systems reacting by inner-sphere paths includes some in which a σ orbital is involved for neither partner [37, 38] (though here the contrast in rates between inner- and outer-sphere paths appears to be small) and a class of systems (39), namely, Pt(II) reacting with Pt(IV), in which two-electron rather than one-electron changes operate.

The examples considered in detail in the preceding sections involve substitution-inert oxidants and oxidized products which are relatively inert to substitution. The condition for success in diagnosis by investigating intermediate products is not that the oxidant and oxidized product be substitution-inert, but only that the rate of electron transfer be no less rapid than the rate of substitution for either of the oxidants. Many systems probably satisfy these conditions, but for the labile ones rather more advanced techniques need to be used. The opportunities in the labile systems for a demonstration of mechanism by flow techniques were first exploited by Sutin and co-workers, and progress in this area is rapidly being made in his laboratory and elsewhere.

An approach [41] which has not been extensively used, but which has the potential of contributing a great deal of information which is otherwise unobtainable on the question of mechanism, is to do experiments at reduced temperatures. The activation energies for most substitution reactions are higher than the activation energies for simple electron-transfer processes of the kind we are considering. By lowering the temperature, many systems which at room temperature are not susceptible to analysis by the method which has been described may yield to it. The approach is obviously of little use for aqueous solutions, unless we consider solutions in water with a high content of electrolyte as aqueous, or unless methods are developed for working with supercooled solutions, and the conclusions reached at low temperature are not necessarily applicable at ordinary temperatures. Nevertheless, the measurements which could be made would be an important contribution as extending definite information on the problem of mechanism, particularly as they apply to reactants which at room temperature undergo substitution more rapidly than they do electron transfer.

Even if the several opportunities which have been described are exploited fully, it may prove to be impossible to assign mechanisms to a number of systems on the basis of direct evidence (that is, by identifying reaction intermediates), and for such systems rate comparisons will have to suffice. However, until the relation between rate patterns and mechanism is established for a large body of data obtained for different systems at different temperatures, in different solvents, and under otherwise varying conditions—establishing the relationship usually requires a number of extrakinetic observations to be made—mechanistic conclusions based on rate measurements can be undependable. It is impossible early in the development of a field of rate behavior to forecast all the complexities which Nature has in store, or to see all the reasonable implications of a particular rate effect which appears to be significant. As a result, conclusions may be based in part on improper exclusions. The history of the present subject provides ample illustration of this statement, and one case will be mentioned in the next chapter. Rate comparisons for halo complexes have been suggested from time to time as diagnostic of mechanism, but some elaboration of the method of comparison is needed before it has much force in argument and the method introduced by Haim [42], which involves investigating the free-energy change attending the exchange of one ligand by another in an activated complex, promises to be useful. A comparison which has always been successful thus far is that for the rate of reduction of an aquo ion compared to the hydroxo complex. It has been mentioned that for outer-sphere reactions, the hydroxo complex is reduced less rapidly than the aquo complex; by the inner-sphere mechanisms, just the reverse is true (in the cobaltpentaammine system, the rate ratio is about $10^7$). Comparisons of this kind are useful only in a qualitative way, and would not be much use in decomposing a reaction which proceeds by parallel paths of different mechanism into the contributions appropriate to each path.

## D. *Other Activation Processes for Inner-Sphere Reactions*

It was assumed earlier that when $Cr^{2+}$ forms an inner-sphere activated complex, a normal coordination position is used. There is no proof of this assumption, but there is some evidence in support of it. The tracer work shows that transfer of oxygen from $Co(NH_3)_5H_2O^{3+}$ to $Cr^{2+}$ is quantitative. If $Cr(H_2O)_7^{2+}$ were produced as the primary product of the reaction, and the seven water molecules were randomized within this complex, loss of water from Co(III) to the solvent would be observed. If a seven-coordinated intermediate is formed, we must assume that at no stage does the entering water molecule become equivalent to one of the other water molecules, and this seems unlikely. If the entering water molecule assumed a normal coordination position on Cr, the quantitative transfer of $H_2O$ as observed is, of course, readily understood.

Experiments with $V(H_2O)_6^{2+}$ have provided proof of the con-clusion that for some reactions a normal coordination position on V(II) is used in forming a bridged· activated complex. In contrast to $Cr(H_2O)_6^{2+}$, which is extremely substitution-labile, substitution on $V(H_2O)_6^{2+}$ is a relatively slow process. The reactions of $V^{2+}$ with $VO^{2+}$ [42], with $Co(NH_3)_5C_2O_4^+$ [43], and with $cis$-$Co(en_2)(N_3)_2^+$ [44] have been shown on the basis of direct evidence to proceed via inner-sphere activated complexes. The first and second of these reactions, a number of others of the same class, and the reactions of $V^{2+}$ with $CrSCN^{2+}$ [45] and with $(NH_3)_5CoN_3^{2+}$ [41] all have kinetic parameters which are remarkably alike, and, moreover, these are consistent with the view that substitution on $V^{2+}$, rather than the electron-transfer act itself, is rate-determining. The definite conclusion for at least a group of $V^{2+}$ reactions has implications which support the assumption that a normal coordination position is used when $Cr^{2+}$ acts in an inner-sphere mechanism. Vanadium(II) com-pared to Cr(II) shows a remarkably high reactivity toward reaction with $Cr(H_2O)_5OAc^{2+}$. This difference can be understood in terms of the fact that the reducing electron for $Cr^{2+}$ is in a $\sigma$ orbital, while that of $V^{2+}$ is in a $\pi$ orbital, but this classification and difference depend on the ligand shared by Cr(III) taking up one of the normal octahedral positions on the reducing agent.

Substitution of the bridging group in a normal coordination position on $V(H_2O)_6^{2+}$ has been emphasized, but it seems alto-gether reasonable that oxidation of $V(H_2O)_6^{2+}$ can take place by an inner-sphere mechanism of an entirely different kind, namely one in which the entering ligand penetrates the coordination sphere of V(II) through the face of an octahedron, thus forming an activated complex, and perhaps even an intermediate of coordination number seven. This kind of mechanism can be illustrated in a simple way for the reaction of atomic chlorine with $V(H_2O)_6^{2+}$. Atomic chlorine can extract one of the $t_{2g}$ electrons of $V^{2+}$, thus creating a vacant orbital which would be bonding toward the approach of a ligand on an octahedral face (the hybrid set $d^3sp^3$ is compatible with the seven-coordinated geometry implied), and it seems quite reasonable to suppose that the system would take advantage of the opportunity offered for minimizing the charge separation attending the electron transfer. If this model is considered reasonable, only a simple elaboration is required to understand the reduction of $Co(NH_3)_5OH^{2+}$ by $V(H_2O)_6^{2+}$ by this kind of mechanism. Proof of this kind of mechanism may be difficult to obtain unless the intermediate seven-coordinate V(III) has a fairly long lifetime.

In contrast to the case discussed, when a one-electron oxidant attacks a reducing agent in which the $t_{2g}$ orbitals are occupied by electron pairs, an antibonding electron is still left in the same orbital after the reducing agent transfers one to the oxidant, so that the ligand given up by the oxidant cannot penetrate the first coordination sphere of the reducing agent. The mechanism, though formally similar to that discussed, would in this case be better described as outer-sphere, and it is clear that the distinction between inner- and outer-sphere processes begins to depend on subtle differences when deviations from orthodox inner-sphere mechanisms are considered. If a two-electron oxidant were to attack a face of a low-spin $t_{2g}^6$ reductant, the situation again would be entirely analogous to that discussed for $V(H_2O)_6^{2+}$.

Not all the reasonable mechanism possibilities have been considered in the foregoing. An important one thus far not mentioned is specific for proton-labile solvents and involves proton transfer accompanying electron transfer. The mechanism has much to recommend it, and evidence in support of it has been adduced, but unambiguous proof that it operates in electron transfer between metal complexes has yet to appear.

## REFERENCES

1. G. T. Seaborg, *Chem. Rev.* **27**, 199 (1940).
2. Symposium on Electron Transfer Processes, *J. Phys. Chem.* **56**, 801–910 (1952).

3. Comment by H. C. Brown following the paper by J. Silverman and R. W. Dodson, in which the influences of anions on rates of electron transfer were discussed, *J. Phys. Chem.* **56**, 896 (1952).

4. W. F. Libby, *J. Phys. Chem.* **56**, 863 (1952), and subsequent discussion.

5. A. M. Zwickel and H. Taube, *J. Am. Chem. Soc.* **83**, 793 (1961).

6. R. B. Jordan, A. M. Sargeson, and H. Taube, *Inorg. Chem.* **5**, 1091 (1966).

7. E. S. Gould, *J. Am. Chem. Soc.* **89**, 977 (1967).

8. N. Sutin, *Ann. Rev. Nucl. Sci.* **12**, 285 (1962).

9. J. Halpern, *Quart. Rev.* **15**, 207 (1961).

10. $Fe^{*3+}(aq) + Fe^{2+}(aq) = Fe^{*2+}(aq) + Fe^{3+}(aq)$    and    $Ru(N^*H_3)_6^{3+} + Ru(NH_3)_6^{2+} = Ru(N^*H_3)_6^{2+} + Ru(NH_3)_6^{3+}$ are self-exchange reactions; $Fe^{3+}(aq) + Ru(NH_3)_6^{2+} = Fe^{2+}(aq) + Ru(NH_3)_6^{3+}$ is a cross reaction.

11. R. J. Campion, C. F. Deck, P. King, Jr., and A. C. Wahl, *Inorg. Chem.* **6**, 672 (1967).

12. R. A. Marcus, *Ann. Rev. Phys. Chem.* **15**, 155 (1964); *J. Phys. Chem.* **67**, 853 (1963).

13. J. F. Endicott and H. Taube, *J. Am. Chem. Soc.* **86**, 1686 (1964).

14. R. J. Campion, N. Purdie, and N. Sutin, *Inorg. Chem.* **3**, 1091 (1964).

15. L. Gjertsen and A. C. Wahl, *J. Am. Chem. Soc.* **81**, 1572 (1959).

16. L. E. Bennett and H. Taube, *Inorg. Chem.* **7**, 254 (1968).

17. R. L. Ward and S. I. Weissman, *J. Am. Chem. Soc.* **79**, 2086 (1957).

18. D. W. Larsen and A. C. Wahl, *J. Chem. Phys.* **43**, 3765 (1965).

19. T. J. Meyer and H. Taube, *Inorg. Chem.* **7**, 2369 (1968).

20. D. R. Stranks, *Discussions Faraday Soc.* **29**, 116 (1960).

21. B. R. Baker, F. Basolo, and H. M. Neumann, *J. Phys. Chem.* **63**, 371 (1959).

22. A. M. Zwickel and H. Taube, *Discussions Faraday Soc.* **29**, 42 (1960).

23. M. T. Barnett, B. M. Craven, H. C. Freeman, N. E. Kine, and J. A. Ibers, *Chem. Commun.* **1966**, 307.

24. H. Taube, H. Myers, and R. L. Rich, *J. Am. Chem. Soc.* **75**, 4118 (1953).

25. H. Taube and H. Myers, *J. Am. Chem. Soc.* **76**, 2103 (1954).

26. J. P. Candlin, J. Halpern, and D. T. Trimm, *J. Am. Chem. Soc.* **86**, 1019 (1964).

27. A. M. Zwickel and H. Taube, *J. Am. Chem. Soc.* **81**, 1288 (1959).

28. R. K. Murmann, F. A. Posey, and H. Taube, *J. Am. Chem. Soc.* **79**, 262 (1957).

29. W. Kruse and H. Taube, **82**, 526 (1960).

30. H. R. Hunt and H. Taube, *J. Phys. Chem.* **63**, 124 (1959).

31. M. Green, K. Schug, and H. Taube, *Inorg. Chem.* **4**, 1184 (1965).

32. A. E. Ogard and H. Taube, *J. Am. Chem. Soc.* **80**, 1083 (1958).

33. A value of $k_H/k_D$ for this equilibrium of 1.4 has been quoted, based on unpublished work of D. Bearcroft and H. Taube. The equilibrium isotope effect for this reaction has been remeasured [R. C. Splinter, J. J. Harris,

and R. S. Tobias, *Inorg. Chem.* **7**, 897 (1968)], using two independent methods which agree on the value 3.0. Unless there is a marked change of the isotopic fractionation factor with $\mu$ (Bearcroft and Taube made their measurements at low $\mu$, while those by Splinter *et al.* were done at $[ClO_4^-] = 0.300\ M$), the more recent measurement must be taken as replacing the earlier ones. In any event, the later measurements would apply to the conditions of the kinetic isotopic fractionation work.

34. Sr. M. J. DeChant and J. B. Hunt, *J. Am. Chem. Soc.* **90**, 3695 (1968).
35. L. E. Orgel, *Inst. Intern. Chim. Solvay, 10ᵉ Conseil Chim. Brussels*, 289 (1956).
36. H. Taube, *J. Chem. Educ.* **45**, 452 (1968).
37. T. W. Newton and F. Baker, *Inorg. Chem.* **3**, 569 (1964).
38. R. J. Campion, T. J. Conocchioli, and N. Sutin, *J. Am. Chem. Soc.* **86**, 4591 (1964).
39. F. Basolo, M. L. Morris, and R. G. Pearson, *Discussions Faraday Soc.* **29**, 80 (1960).
40. See, for example, A. Haim and N. Sutin, *J. Am. Chem. Soc.* **88**, 5343 (1966).
41. M. Ardon, J. Levitan, and H. Taube, *J. Am. Chem. Soc.* **84**, 872 (1962).
42. T. W. Newton and F. B. Baker, *J. Phys. Chem.* **68**, 228 (1964).
43. H. J. Price and H. Taube, *Inorg. Chem.* **7**, 1 (1968).
44. J. H. Espenson, *J. Am. Chem. Soc.* **89**, 1276 (1967).
45. B. R. Baker, M. Orhanovic, and N. Sutin, *J. Am. Chem. Soc.* **89**, 722 (1967).
46. J. P. Candlin, J. Halpern, and D. L. Trimm, *J. Am. Chem. Soc.* **86**, 1019 (1964).

# III

## SOME ASPECTS OF LIGAND EFFECTS IN ELECTRON-TRANSFER REACTIONS

### I. Introduction

Trying to understand the effect [1] which ions such as Cl⁻ have in promoting isotopic exchange between $Fe^{2+}$ and $Fe^{3+}$ provided a great deal of the incentive for the early research on the mechanism of electron-transfer reactions, and the interest in the effect of ligands on the rate and mechanism these reactions has grown as more has been learned about the subject. The proof that metal complexes can undergo electron transfer by inner-sphere mechanisms (as well as by outer-sphere mechanisms) led to a focusing of effort in the field by drawing attention to the special role which the bridging ligands play. To understand the mechanism of electron transfer through bridging ligands is an important goal of current research in this general field. In considering mechanism in this context, we have in mind trying to distinguish electron transfer by a "chemical" mechanism from a "resonance" mechanism [2]. The term "chemical" as used here refers to a process in which either the oxidizing metal ion is strong enough to oxidize, or the reducing metal ion is strong enough to reduce the ligand which the two share, the electron defect or the excess electron then being passed on to the reducing or oxidizing metal ion, respectively. When the chemical mechanism operates in the electron-excess case, transfer takes place by "hopping," and the electron passes from a well-defined bound state in the reducing agent to another in the ligand, whereupon it eventually passes to the oxidizing center and is trapped there. In contrast, when resonance transfer occurs, the electron is assumed to pass directly from a bound state on the reducing agent to another on the oxidizing metal ion without occupying a bound state on the mediating ligand. Even when the electron in transferring does not occupy a well-defined bound state on the ligand, the rate of electron transfer can be very sensitive to the nature of the ligand. Since a binuclear transition state is formed, in considering how ligand properties affect the rate we must, at the very least, reckon with the different capacities of ligands to stabilize binuclear combinations; but beyond this, the probability of transfer of the electron will be sensitive to the presence on the ligand of low-lying unoccupied orbitals. The energy levels on the ligand determine the height of the barrier which the electron must penetrate and thereby affect the probability of transfer.

The later discussion will show that differences in the behavior of systems can be expected at least in the extremes of mechanism as they have been described. The effects noted to date which are associated with the extremes in mechanism are largely a matter of rates of reaction, the reactivity pattern for a series of selected systems being sensitive to the kind of mechanism operating. But only after a large amount of information is accumulated and understood do conclusions about mechanism based on rate differences alone become at all firm. The history of enquiry along the lines being discussed is still quite short, and experience with the systems is limited. Thus, most of the conclusions about mechanism offered, though they may appear to be reasonable in the light of our present level of understanding, are to be regarded as tentative.

## II. SIMPLE LIGANDS

The halides, because they form complexes with a large number of different oxidizing metal ions, and because of their simple structure, appear to offer a good opportunity for defining the role of the bridging ligand in electron-transfer reactions, and thus have figured prominently in studies on bridging ligand effects. When halide complexes are reduced by cationic reducing agents which act by outer-sphere mechanisms, in every system encountered thus far, the rate of reduction increases as the size of the halide increases [3, 4]. For inner-sphere reactions, no such simple generalization is possible; this assertion is borne out by data obtained with $Cr^{2+}$ as reducing agent and summarized in Table III-1. The data for $Cr^{2+}$ were selected because of the

TABLE III-1

COMPARISON OF RATES FOR THE REDUCTION OF HALO COMPLEXES BY $Cr^{2+a}$

| Complex | $F^-$ $M^{-1} sec^{-1}$ | $Cl^-$ $M^{-1} sec^{-1}$ | $Br^-$ $M^{-1} sec^{-1}$ | $I^-$ $M^{-1} sec^{-1}$ | Reference |
|---|---|---|---|---|---|
| $Cr(NH_3)_5^{3+}$ | $2.7 \times 10^{-4}$ | $5.1 \times 10^{-2}$ | $0.32$ | — | [6] |
| $Cr(H_2O)_5^{3+}$ | $2.6 \times 10^{-3}$ | $9$ | $> 60$ | — | [7] |
| $Co(NH_3)_5^{3+}$ | $2.5 \pm 0.5 \times 10^5$ | $6 \pm 1 \times 10^5$ | $1.4 \pm 0.4 \times 10^6$ | $3 \pm 1 \times 10^6$ | [8] |
| $Ru(NH_3)_5^{3+}$ | — | $3.5 \times 10^4$ | $2.2 \times 10^3$ | $< 5 \times 10^2$ | [5] |

[a] The data pertain to 25° except for $Cr(H_2O)_5^{3+}$ at 0°. Ionic strengths are 1.0, 1.0, 0.10, and 0.10 $M$ for $Cr(NH_3)_5^{3+}$, $Cr(H_2O)_5^{3+}$, $Co(NH_3)_5^{3+}$, and $Ru(NH_3)_5^{3+}$, respectively.

reducing agents which have been rather intensively studied—e.g., $Eu^{2+}$, $Cr^{2+}$, $V^{2+}$, and $Fe^{2+}$—it is the one for which we can, with the greatest confidence, reach the conclusion that an inner-sphere reaction mechanism operates in a particular instance.

The reversal of the trend in rates which occurs when $Cr(NH_3)_5^{3+}$ complexes are compared with those of $Ru(NH_3)_5^{3+}$ is quite striking. On the basis of precedent established for the abstraction by radicals of halides from carbon compounds, it is the Ru complexes that show "unusual" behavior, but this behavior seems unusual only because the relevant experience is very limited.

The reversal of the trend in rates exhibited in Table III-1 shows clearly that the bare data as recorded there cannot be used as a criterion for classifying mechanism except in limited situations. For all outer-sphere reductions thus far studied, the rate for the fluoro oxidant is less than that for the iodo and thus, when the reverse order is recorded, it probably means that an inner-sphere path makes an important contribution to the reaction. This trend has been observed (see Table III-2) in the reduction by $Eu^{2+}$ [9] and by $Fe^{2+}$ [10, 11] of complexes of the family $Co(NH_3)_5X^{2+}$, and the meaning indicated above has been attached to the observations. However, when the trend in rates $I^- > F^-$ is observed, it is not immediately obvious how the data can be used to diagnose mechanisms. A recent effort [12] which has been made in this direction and which does give greater insight into the meaning of the data is herewith outlined and commented on.

The importance in determining the rate in an inner-sphere reaction of the strength of the bond which the transferring ligand forms compared to that which it loses has been variously pointed out [9, 11], as has the arbitrary character [13] of casting the rate data for the reaction of a reducing agent $G^{2+}$ acting on an oxidizing agent $J^{3+}$ with $X^-$ present in the form $(G^{2+})$ $(JX^{2+})$ rather than as $(GX^+)(J^{3+})$ or $(G^{2+})(J^{3+})(X^-)$. The stricture as to the form of the rate law applies particularly to cases in which both ions undergo substitution rapidly compared to electron transfer, so that the position of $X^-$ in the activated complex cannot be determined from kinetic data. To compare the relative efficiency of $I^-$ and $F^-$ in facilitating electron transfer between $Co(NH_3)_5^{3+}$ and $Cr^{2+}$ (choosing now a specific example for illustration) Haim suggests using the equilibrium constant $Q_{I,F}$ for the formal substitution reaction

$$[(NH_3)_5CoFCr^{4+}]^{\ddagger} + I^- = [(NH_3)_5CoICr^{4+}]^{\ddagger} + F^- \quad \text{(III-1)}$$

It should be noted that in reaction (III-1) we are dealing with an equilibrium constant involving not stable species, but activated complexes. The value of $Q_{I,F}$ can be calculated from the rate and equilibrium constants of the reactions shown below:

$$Co(NH_3)_5F^{2+} + Cr^{2+} = [(NH_3)_5CoFCr^{4+}]^{\ddagger}, \qquad k_F \quad \text{(III-2)}$$

$$Co(NH_3)_5I^{2+} + Cr^{2+} = [(NH_3)_5CoICr^{4+}]^{\ddagger}, \qquad k_I \quad \text{(III-3)}$$

$$Co(NH_3)_5OH_2^{3+} + F^- = Co(NH_3)_5F^{2+} + H_2O, \qquad Q_F \quad \text{(III-4)}$$

$$Co(NH_3)_5OH_2^{3+} + I^- = Co(NH_3)_5I^{2+} + H_2O, \qquad Q_I \quad \text{(III-5)}$$

with $Q_{I,F} = k_I Q_I / k_F Q_F$. The meaning of the quotient $Q_{I,F}$ perhaps becomes intuitively clearer if we recognize that it is equivalent to comparing the specific rates in terms of the net activation processes written showing the Co(III) complexes in some common form, chosen here for convenience as $Co(NH_3)_5OH_2^{3+}$:

$$(NH_3)_5CoOH_2^{3+} + I^- + Cr^{2+} \rightarrow P, \qquad k_I' \quad \text{(III-6)}$$

$$(NH_3)_5CoOH_2^{3+} + F^- + Cr^{2+} \rightarrow P', \qquad k_F' \quad \text{(III-7)}$$

Now $k_I' = Q_I k_I$ and $k_F' = Q_F k_F$ so that $Q_{I,F} = k_I'/k_F'$.

When $k_I'$ and $k_F'$ are compared, the iodide complex of Co(III) is not given the advantage of a lower stability compared to the fluoride, as is the case when $k_I$ and $k_F$ are compared, and in using $k_I'$ and $k_F'$ the stabilities of the activated complexes are related to a common state of the reactants. This point is illustrated in Fig. III-1, which also shows other intermediate states in relation to the initial state in which the halide bonds to neither reaction partner.

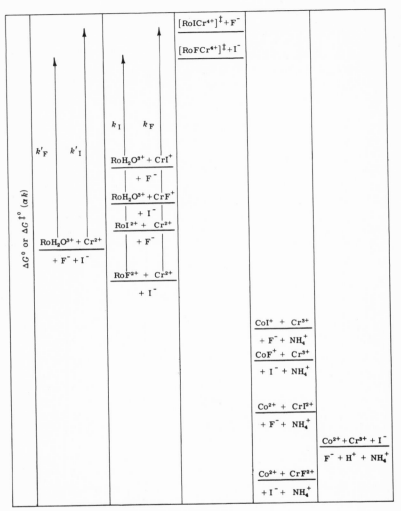

FIG. III-1. Profile of $\Delta G^0$ in the system

$$Co(NH_3)_5^{3+}(\equiv Ro^{3+}) + I^- + F^- + Cr^{2+}.$$

The species $H_2O$ is not shown except when it is in the coordination sphere of $RoH_2O^{3+}$. The energy of the species $CrI^+$ relative to $CrF^+$ and $CoI^+$ relative to $CoF^+$ is not known, but it has been assumed that the fluoro complex is more stable than the iodo. No attempt has been made to represent the vertical scale accurately.

There is little to object to in the analysis thus far. Comparing the rates in terms of net activation processes such as (III-6) and (III-7) has a clear meaning, and has the advantage that the rate of a reaction such as

$$Co(NH_3)_6^{3+} + Cr^{2+} + X^- \rightarrow \qquad \text{(III-8)}$$

which proceeds by an outer-sphere activated complex, and for which there is no alternative [14] except to express the rate by the function $[Co(NH_3)_6^{3+}][Cr^{2+}][X^-]$, is now commensurable with the rate for an inner-sphere mechanism. However, in coming this far, we have at best succeeded in formulating the problem, but have yet to answer the basic questions as to the meaning of the trends in the values of $Q_{X,Y}$. In this connection, it should be mentioned that the three categories which were defined by Haim dealing with these trends, and which, on the basis of a limited field of data, he suggested as criteria for the classification of mechanism, appear not to be internally consistent when they are generalized. The three categories he proposes are (1) inner-sphere reductions by class A, or "hard" metal centers ($Q_{I,F} < 1$); (2) outer-sphere reductions ($Q_{I,F} > 1$); (3) inner-sphere reductions by class B, or "soft" metal ions ($Q_{I,F} > 1$). Let us consider a case in which an oxidizing center reducible to a class B metal ion reacts with a class A metal ion. According to the categories suggested by Haim, $Q_{I,F} < 1$ for the forward reaction. Now, when the net changes are expressed as

$$M^{3+} + X^- + Cr^{2+} = M^{2+} + X^- + Cr^{3+} \qquad \text{(III-9)}$$

$K_{eq}$ is independent of $X^-$, and therefore for the reverse reaction $Q_{I,F}$ must also be less than 1 (if the equilibrium constant is unaltered by replacing $F^-$ by $I^-$, then if the forward reaction proceeds more rapidly in the presence of $F^-$, so must the reverse). But since $M^{2+}$ is a class B metal ion, the reverse reaction belongs in category 3, for which $Q_{I,F}$ is asserted to be greater than 1. The difficulty with the categories 1 and 3 mentioned is that they are based on the role of the reducing agent, even though according to the formulation embodied in reactions (III-6) and (III-7) the systems are symmetrical with respect to oxidant and reductant. Thus it follows that if the quality of the reductant is a factor, that of the oxidant must also be considered.

The criticism offered applies only to the criteria of mechanism offered by Haim, and not to the suggestion as to what is a useful way of comparing the rate data. Accepting this suggestion, an attempt will be made in the following to indicate the form the questions about the role of the bridging ligand take when the energies of the activated complexes for a given oxidizing agent in combination with a given reducing agent are referred to a common basis. Particularly instructive are the observations for a reaction such as

$$Co(NH_3)_5X^{2+} + Ru(NH_3)_6^{2+} \rightarrow \qquad (III-10)$$

which takes place by an outer-sphere mechanism. On being recast, the reaction takes the form

$$Co(NH_3)_5OH_2^{3+} + X^- + Ru(NH_3)_6^{2+} \rightarrow \qquad (III-10a)$$

From the relevant data, $Q_{I,Cl} = (k_I'/k_{Cl}')$ is calculated to be $3.5 \times 10^2$, i.e., the "affinity" of the activated complex for $I^-$ exceeds that for $Cl^-$ by the factor $3 \times 10^2$. The point of special interest is to find the factor or factors producing the differences in affinity. For the cobalt–ammine complex in its ground state, the affinity ratio for $I^-$ relative to $Cl^-$ is 0.1 [15]. In the activated complex there is a bond between Co and $X^-$, and the state of cobalt is expected to be between that of $Co^{3+}$ and $Co^{2+}$, so that the affinity ratio in terms of bond formation is expected to be greater than 0.1, but still less than 1. (It is reasonable to suppose that the affinity order for $Co^{2+}$ with the halides will be $F^- > Cl^- > Br^- > I^-$.) Thus the greater affinity which the activated complex has for $I^-$ over $F^-$ cannot be ascribed to $Co-X$ bond formation in the activated complex. The halide is prevented by the substitution inertia of $Ru(NH_3)_6^{2+}$ from making an inner-sphere bond to Ru, and therefore it is unlikely that

Ru$-$X bond formation in the activated complex accounts for the affinity trend. We must conclude that a factor other than bond formation contributes to the "affinities" which the activated complexes show for the halide ions. This factor is important enough to overcome the effect of partial bond formation, and it is likely that the factor in question is the "permeability" of the halide ion to electron or electron-defect flow [13].

Both bond formation and permeability to electron flow are, in principle, factors affecting the affinity of activated complexes for ligands also when the ligands bridge both partners, but now the resolution into the two components is rendered more difficult. It would be a great simplification of the subject if the "permeability" were constant from one situation to another, but it is not likely that this will prove to be the case. The difference in rate between Cl$^-$ and I$^-$ as recorded above is large when the halide can occupy a position in the coordination sphere of the oxidant. However, when it is excluded from the coordination spheres of both reactants [3,16], the difference in rate as the halide is varied becomes much less, and is now so small that the effect of the permeability on rate is obscured by other factors. The comparison of the two cases does not provide a basis for extrapolating to the inner-sphere activated complex, as the third example involving an outer-sphere activated complex will show.

The rate of reaction between $Co(NH_3)_6^{3+}$ and $V(H_2O)_6^{2+}$ is extremely sensitive to the presence of complexing agents in solution [17], and when the rate is expressed in the form

$$k[Co(NH_3)_6^{3+}][V(H_2O)_6^{2+}][X^-] \qquad \text{(III-11)}$$

$k_F' = 2.6 \times 10^3\, k_{Cl}'$. In this outer-sphere activated complex, in which the halide can make a bond only to the reducing agent, the trend is opposite to that observed when the halide can make a bond only to the oxidant. If the permeability of the halide is a factor in the activated complex corresponding to rate law (III-11) [and it need not be because the halide ion may not be presented in the configuration $(NH_3)_5CoNH_3 \cdot XV(H_2O)_5$], this is more than compensated for by the effect of bond formation. Vanadium in the activated complex is in a state between $2+$ and $3+$, and the effect of the bond formation would lead to the affinity order $F^- > I^-$, as observed. The halide ion may play quite a different role when $Co(NH_3)_5X^{2+}$ reacts with $Ru(NH_3)_6^{2+}$ than it does when $Co(NH_3)_6^{3+}$ reacts with $V(H_2O)_6^{2+}$ in the presence of halide. For the former system, but not the latter, electron transfer through the halide is suggested.

Even if we are not in position to understand the ligand effects for inner-sphere activated complexes, it is instructive to discuss some of the observations when they are presented in the form illustrated by Eq. (III-9). In the reactants, $X^-$ does not make a bond to either, so that the rate difference observed on replacing $X^-$ by $X'^-$ can be attributed to differences in the affinities of the activated complexes for $X^-$ and $X'^-$. The reaction of $(H_2O)_5CrX^{2+}$ with $Cr^{2+}$ is particularly interesting in the present context, because here the net free-energy change (except for that associated with isotopic redistribution) is zero. From the results of Ball and King [7] and the data on the affinities of $Cr^{3+}$ for $F^-$ and $Cl^-$, $Q_{Cl,F}$ can be calculated to be about $2 \times 10^{-3}$. The activated complex thus shows a greater affinity for $F^-$ than for $Cl^-$, but this discrimination is very much less [18, 19] than that exerted by $Cr^{3+}$. The result can be taken to indicate that the $X-Cr$ bonds in the activated complex have been so much weakened compared to the initial state that the discrimination between the halides is very much reduced, or if substantial bonding is retained, it would suggest that permeability remains an important factor, which to some extent offsets the effect of bonding. For the reactions of $Co(NH_3)_5^{3+}$ with $Cr^{2+}$, $Q_{Cl,F} = 0.1$ while $Q_{Br,Cl} = 0.8$; for the reaction of $Ru(NH_3)_5^{3+}$ with $Cr^{2+}$, $Q_{Br,Cl} = 0.03$ [20]. These comparisons introduce a new complication, and a resolution of the questions raised by the observations is even less to be expected. When $Co(NH_3)_5^{3+}$ complexes are compared to those of $Cr(H_2O)_5^{3+}$, acceptor orbitals of the same symmetry are involved, but the former class of complexes are much stronger oxidants. The driving force for the reduction of $Ru(NH_3)_5^{3+}$ is about the same as for the reduction of $Co(NH_3)_5^{3+}$, but the acceptor orbital for the former has $\pi$ while for the latter it has $\sigma$ symmetry. The orbital symmetry at the very least has this indirect effect: when the acceptor orbital has $\pi$ symmetry, it is nonbonding, and the incoming electron is accommodated without much change in metal-bridging ligand bond distance, while when it has $\sigma$ symmetry, the incoming electron is antibonding, and considerable stretching of the metal-bridging ligand bond must take place. There may also be direct effects of the orbital symmetry in these simple ligands of the kind which will be documented for ligands having conjugated bond systems, but these have not been clearly demonstrated.

## III. Activation Parameters

Thus far we have dealt with rate comparisons at a fixed temperature, but the temperature is obviously an important variable that needs to be taken into account. A theory which is framed to fit data at a fixed temperature may not have the capacity to account for a change in reactivity pattern as the temperature changes, and if it were lacking in this respect, it would have to be judged incomplete, if not actually altogether wrong. But the rate data as a function of temperature, quite apart from their real use in providing a stringent test of theory, have a use also in leading to values of the quantitatively descriptive kinetic parameters $\Delta H^{\ddagger}$ and $\Delta S^{\ddagger}$. Of these, $\Delta S^{\ddagger}$, having as it does a more direct connection with the molecularity, structure, and charge of the activated complex, is regarded as being the more directly useful in terms of providing insight into reaction mechanism. In particular, the hope has been expressed from time to time that the values of $\Delta S^{\ddagger}$ would be helpful in distinguishing between inner- and outer-sphere mechanisms. Enough data of sufficient accuracy have by now been collected to make it worthwhile to investigate this point for reactions of halo complexes. Most of the relevant data are summarized in Table III-2.

The conclusion that the reactions of $Cr^{2+}$ dealt with in Table III-2 take place by inner-sphere mechanisms is dependable because in every case, except for the reaction of $Ru(NH_3)_5Br^{2+}$, it is based on the identification of intermediate products, and

TABLE III-2

ACTIVATION PARAMETERS FOR THE REDUCTION OF HALO COMPLEXES

| Reaction | Mechanism[a] | $\mu$ | $k$ ($M^{-1} sec^{-1}$ at 25°) | $\Delta H^{\ddagger}$ (kcal/mole) | $\Delta S^{\ddagger}$ (e.u.) | Reference |
|---|---|---|---|---|---|---|
| $Cr^{2+} + Cr(NH_3)_5F^{2+}$ | i.-s. | 1.00 | $2.7 \times 10^{-4}$ | 13.4 | —30 | [6] |
| $Cr^{2+} + Cr(NH_3)_5Cl^{2+}$ | i.-s. | 1.00 | $5.1 \times 10^{-2}$ | 11.1 | —28 | [6][b] |
| $Cr^{2+} + Cr(NH_3)_5Br^{2+}$ | i.-s. | 1.00 | $3.2 \times 10^{-1}$ | 8.5 | —33 | [6] |
| $Cr^{2+} + Cr(H_2O)_5F^{2+}$ | i.-s. | 1.0 | $2.6 \times 10^{-3}$ | 13.7 | —20 | [7] |
| $Fe^{2+} + Co(NH_3)_5F^{2+}$ | — | 1.0 | $6.6 \times 10^{-3}$ | 13.7 | —23 | [10][c] |
| $Fe^{2+} + Co(NH_3)_5Cl^{2+}$ | — | 1.0 | $1.4 \times 10^{-3}$ | 12.5 | —30 | [10][c] |
| $Fe^{2+} + Co(NH_3)_5Br^{2+}$ | — | 1.0 | $0.73 \times 10^{-3}$ | 13.3 | —28 | [10][c] |
| $Fe^{2+} + FeF^{2+}$ | — | 0.50 | 9.7 | 9.1 | —21 | [21] |
| $Fe^{2+} + FeCl^{2+}$ | — | 0.55 | 9.7 | 8.8 | —24 | [1] |
| $V^{2+} + Co(NH_3)_5Br^{2+}$ | — | 1.00 | 25 | 9.1 | —22 | [9] |
| $Eu^{2+} + Co(NH_3)_5Cl^{2+}$ | — | 1.00 | $3.9 \times 10^2$ | 5.0 | —30 | [9] |
| $Eu^{2+} + Co(NH_3)_5Br^{2+}$ | — | 1.00 | $2.5 \times 10^2$ | 4.7 | —32 | [9] |
| $Cr^{2+} + Ru(NH_3)_5Cl^{2+}$ | — | 0.10 | $3.5 \times 10^4$ | 1.3 | —33 | [5] |
| $Cr^{2+} + Ru(NH_3)_5Br^{2+}$ | i.-s. | 0.10 | $2.2 \times 10^3$ | 2.8 | —34 | [5] |
| $V^{2+} + Ru(NH_3)_5Cl^{2+}$ | o.-s. | 0.10 | $3.0 \times 10^3$ | 3.8 | —30 | [5] |
| $V^{2+} + Ru(NH_3)_5Br^{2+}$ | o.-s. | 0.10 | $5.1 \times 10^3$ | 2.8 | —34 | [5] |

[a] i.-s. = inner-sphere, o.-s. = outer-sphere.

[b] The value reported in [6] was miscalculated.

[c] The data reported in [11], also dealing with this reaction, were obtained at $\mu = 1.7$ rather than $\mu = 1.0$. The values of $\Delta S^{\ddagger}$ for X = F, Cl, and Br at $\mu = 1.7$ are given as —23, —23, and —20 e.u., respectively.

even for $Ru(NH_3)_5Br^{2+}$ the indirect evidence in support of the conclusion is so strong as to be tantamount to proof. The conclusion as to the mechanism of the two reactions shown of Ru(III) complexes with $V(H_2O)_6^{2+}$ is almost as dependable as it is for the reactions of $Cr^{2+}$, but the basis for the conclusion is somewhat different. Substitution in $V(H_2O)_6^{2+}$ takes place so slowly that the upper limit for the rate of the bimolecular substitution process is of the order of $10^2 M^{-1} sec^{-1}$, which is much less than the observed rate of the redox reactions. Thus, unless some new effects are involved, it seems safe to conclude that the redox reaction does not involve substitution into a normal coordination position for $V(H_2O)_6^{2+}$. Despite the difference in mechanism, the values of $\Delta S^\ddagger$ for the reactions of $Cr^{2+}$ and $V(H_2O)_6^{2+}$ with Ru(III) complexes are almost the same, and we must conclude that a superficial use of the values of $\Delta S^\ddagger$ does not lead to a safe classification of mechanism.

Perhaps the most remarkable feature of the data shown in Table III-2 is that so many of the values of $\Delta S^\ddagger$ lie in the range $-31 \pm 3$, regardless of the reactant. A few are noticeably smaller, but these are rare enough that it is natural to look for some special effects in these cases, with the possibility of experimental error being included among the special causes, at least in certain of the reactions.

The systems have in common the property that dipositive cations are being brought together to form the activated complexes, and we need to examine the proposition that the values of $\Delta S^\ddagger$ simply reflect the increased involvement of the solvent associated with the process

$$M^{2+} + M'^{2+} \rightarrow [MM']^{4+} \qquad \text{(III-12)}$$

The data recently reported [22] for the reaction of $V(H_2O)_6^{2+}$ with complexes of the type $[Co(NH_3)_5O_2CCOY]^{2+}$ are especially significant in this context. Here, too, the *net* activation process (which determines the value of $\Delta S^\ddagger$ irrespective of the details of the path leading to the activated complex) is of the charge type of reaction (III-12), but the values of $\Delta S^{\ddagger}$ are in the range $-14$ to $-17$ e.u. But the rate-determining steps in the reactions of $V(H_2O)_6^{2+}$ discussed involve substitution in the reducing complex, not electron transfer. The comparison of the values of $\Delta S^{\ddagger}$ when substitution is rate-controlling with those when electron transfer is rate-controlling revives the idea [23] that an abnormally large entropy decrease is associated with the latter kind of reaction. This, in turn, invites speculation that the redox reactions often feature low values of the transmission coefficient, as defined in the Eyring formulation of rates of reaction.

Though it has been emphasized that most of the reactions shown in Table III-2 have values of $\Delta S^{\ddagger}$ in the range of $-30$ e.u., some of the lower values recorded are undoubtedly real, and it is important to understand the reason for the differences. The data are, however, still too few in number to make it profitable to record extensive speculation on the subject. For example, on the basis of the present data, one may ask if low numerical values of $\Delta S^{\ddagger}$ are typical of reactions of $Fe^{2+}$; this conclusion would be justified if the entropies reported by Diebler and Taube [11] were accepted in place of those entered in Table III-2. The difference between the two sets of values of $\Delta S^{\ddagger}$ may be a result of the difference in ionic strength, but it is likely that the differences are in part attributable to experimental inaccuracies. Accurate values of entropies of activation for these and other systems are needed, including reactions proceeding by different kinds of outer-sphere activated complexes. Though the experiments leading to these are unlikely to be exciting in themselves, the data are important and can prepare the way for a basic advance in understanding the subject.


## IV. The Mechanism of Electron Transfer

The questions concerning the mechanism of electron transfer, which in our present state of knowledge appear to be operational, were introduced in a previous section. For some of the systems described thus far, an unambiguous decision can be reached with respect to the questions which were raised, and a decision seems to be particularly easy for the reaction of $Cr^{2+}$ with $Cr(H_2O)_5F^{2+}$ or $Co(NH_3)_5F^{2+}$. The two systems are similar and, with respect to the particular issue under consideration, the discussion of one applies also to the other, so only the former will be dealt with. The chemical mechanism would require the system to pass through the state

$$[(H_2O)_5Cr^{(III)} \cdot F^{2-} \cdot Cr^{(III)}(H_2O)_5]^{4+}$$

or

$$[(H_2O)_5Cr^{(II)} \cdot F \cdot Cr^{(II)}(H_2O)_5]^{4+}$$

The reduction of $F^-$ by $Cr(II)$, as indicated in the first alternative, is endoergic by several electron volts [note that charge transfer absorption in $Cr(H_2O)_6^{2+}$, which undoubtedly takes place at lower energy than in $Cr(H_2O)_5F^+$, begins in the ultraviolet region of the spectrum] and in the second again by several electron volts ($Cr^{3+}$ is a much weaker oxidant than $F^-$). With such barriers to overcome, very high activation energies would be expected for reaction by either of the intermediates and, in consequence, rates very *much* smaller than those observed would be expected. Whatever the mechanism of electron transfer in this system, it is certainly not the simple chemical one, and mainly for want of a more detailed description, we will describe the mechanism which does operate as resonance transfer.

It is quite likely that all of the reactions considered in detail thus far in this chapter take place by resonance transfer. All of them involve electron transfer through single atoms or ions, and in every case the energy barrier to a chemical mechanism seems to be prohibitively high. Observations which have been made with complexes containing large ligands do, however, suggest that a chemical mechanism operates in certain of these systems.

Fairly late [24, 25, 26] in the history of the subject of electron transfer through conjugated bond systems, it became apparent that for facile electron transfer from $Cr^{2+}$ to Co(III) through a ligand, a conjugated bond system connecting the two metal centers is not a sufficient condition, though it may possibly be a necessary one. At the least, one other parameter needs to be introduced, and this parameter is called the "reducibility" of the ligand. For present purposes, reducibility will be defined as the tendency for the reaction

$$L + e = L^- \tag{III-13}$$

to take place. The definition is straightforward, but the measurement of reducibility according to this definition is by no means straightforward in every case, and the term will be used rather loosely. The criterion for reducibility as it is to be used here is the ease of net reduction by $Cr^{2+}$, but it is admitted that there need be no simple relation between the rate of net reduction, involving a two-electron change, and the equilibrium implied by equation (III-13), involving a one-electron change, or the availability of a low-lying unoccupied orbital as deduced from the spectrum or from the electronic structure.

The need to invoke a criterion in addition to conjugation is illustrated by comparing [24] the rate of reduction by $Cr^{2+}$ of

$$\left[ (NH_3)_5Co-O-\underset{\underset{O}{\|}}{C} - \bigcirc \right]^{2+}$$

with that of

$$\left[ (NH_3)_5Co-O-\underset{\underset{O}{\|}}{C} - \bigcirc NCH_3 \right]^{3+}$$

In both cases the ligand is transferred to chromium, and in both, therefore, attack takes place at the carboxyl function. The local environment at each carboxyl is the same, and steric effects are therefore much the same for both complexes. The effect of charge alone would be expected to lower the rate of reduction of the pyridine complex. The rates of reduction are, however, observed to be 0.15 and 1.3 $M^{-1}$ sec$^{-1}$, respectively, at 25°. Esters of the pyridinium cation have been shown [27] to be readily reducible, but this is by no means true of benzoic acid, and, in view of the nature of the process we are considering, it seems reasonable, therefore, to ascribe the difference in rates to the difference in reducibility of the ligand.

Another comparison which illustrates the importance of ligand reducibility involves the reductions of

$$\left[ (NH_3)_5Co-O-\underset{\underset{O}{\|}}{C}-CHO \right]^{2+}$$

and of

$$\left[ (NH_3)_5CoO-\underset{\underset{O}{\|}}{C}-CH_2OH \right]^{2+}$$

by $Cr^{2+}$. The specific rate for the former reaction [28] is 3.1 at 25° and $\mu = 1.0$. This rate is higher than that observed for the acetate complex [29] (which is 0.34 under the same conditions) and the greater rate for the glycolate complex can reasonably be ascribed to its chelating capacity. Now the chelating capacity of the glyoxylate ligand is undoubtedly less than that of the glycolate, yet the rate of reduction of the glyoxylate complex exceeds $7 \times 10^3 M^{-1}$ sec$^{-1}$ [22]. The ligands glycolate and glyoxylate again differ in reducibility—the latter, but not the former, reacts rapidly with $Cr^{2+}$—and again it seems reasonable to attribute the rate difference observed for the two complexes to a difference in reducibility of the ligands.

A relationship between reducibility of ligands and rates of electron transfer does not help to distinguish between the two mechanisms of electron transfer being considered, at least on the qualitative level. By either mechanism, the lower the energy of an unoccupied orbital, the greater should be the rate of reaction. For the chemical mechanism, this relationship is obvious; for the resonance mechanism it is perhaps less obvious, but is equally valid (at least qualitatively) because the lower the energy of an unoccupied orbital, the smaller the energy barrier to electron "tunneling." To distinguish between the mechanisms we turn to the point that in the chemical mechanism the acceptor center is not necessarily involved in the rate-determining step, while in the tunneling mechanism it necessarily is.

For the chemical mechanism, the following sequence of steps can be considered:

$$Co^{(III)}L + Cr^{2+} \rightleftharpoons Co^{(III)} \cdot L^- \cdot Cr^{(III)} \qquad \text{(III-14)}$$

$$Co^{(III)} \cdot L^- \cdot Cr^{(III)} \rightarrow products \qquad \text{(III-15)}$$

According to the mechanism, the rate of formation of products is given by the expression

$$\frac{k_{15} k_{14}}{k_{15} + k_{-14}} [Co^{(III)}L][Cr^{2+}]$$

The nature of the acceptor will not markedly affect the rates of reactions (III-14) except for secondary influences arising from differences in charge, but is expected markedly to affect the rate of reaction (III-15). In the general case, since $k_{15}$ appears in the rate expression, the rate of the overall reaction will depend on the properties of the acceptor center. However, if the situation is realized that $k_{15}$ is greater than $k_{-14}$, or is of the same order of magnitude, the rate of reaction will be rather insensitive to the nature of the acceptor group. This situation appears to be realized [30] in the reactions of $Cr^{2+}$ with

$$\left[ (NH_3)_5 CoN \bigcirc - C_{NH_2}^{O} \right]^{3+} \quad and \ with \quad \left[ (H_2O)_5 CrN \bigcirc - C_{NH_2}^{O} \right]^{3+}$$

The specific rate of reaction of the former complex with $Cr^{2+}$ is 17.6 $M^{-1}$ sec$^{-1}$ (25°, $\mu = 1.0$) while for the latter it is 1.8 $M^{-1}$ sec$^{-1}$. When ions such as $F^-$, $OH^-$, or $Cl^-$ are the bridging groups, the rates at which the pentaamminecobalt complexes react with $Cr^{2+}$ are at least $10^5$ greater than the rates at which the pentaquochromium complexes react; a relationship which, because of the very much greater driving force for the reduction of the Co(III) complexes, is not unexpected. In contrast to the simple bridging groups, isonicotinamide is readily reducible; this fact, together with the rate relationship referred to above, suggests that the redox mechanism corresponds to that embodied in reactions (III-14) and (III-15). Electron transfer from $Cr^{2+}$ to the bridging group is virtually rate-determining, and thus a chemical mechanism for electron transfer in these systems is indicated.

An additional result obtained with

as oxidizing agent [31] shows that the conclusion reached above is not applicable to all systems and, moreover, suggests that there is an interesting relation between the mechanism of electron transfer and the electronic structure. The ruthenium complex is only slightly more powerfully oxidizing than the cobalt complex, but the rate of its reduction by $Cr^{2+}$ under the conditions applying to the experiments with the Co(III) and Cr(III) complexes exceeds $5 \times 10^5 \, M^{-1}$ sec$^{-1}$.

The difference in the rates of reduction of the Cr(III) and Co(III) complexes on the one hand and the Ru(III) complex on the other may be a consequence of the fact that the acceptor orbitals for the first pair have $\sigma$ symmetry with respect to the ligands, but for the latter ion the acceptor orbital has $\pi$ symmetry. Effective overlap between the acceptor orbital of Cr(III) and

Co(III) and the $\pi$ orbital of the ligand can take place only in a distorted configuration of the coordination sphere of the oxidizing ion. The acceptor orbital on Ru(III) has the proper symmetry to mix with a low-lying unoccupied orbital of the ligand. The rate difference does not necessarily imply a fundamental difference in mechanism between Co(III) as an acceptor and Ru(III) as acceptor, but can perhaps be understood in terms of the effect of the two centers on the energy of the acceptor orbital. In the case of Co(III), this can be purely a ligand orbital, but in the case of Ru(III), it can be a mixed ligand-metal ion orbital, and in the latter case, electron transfer to the ligand is not distinguishable from electron transfer to the metal.

## V. CONCLUSION

The preceding discussion has indicated some of the limits of our knowledge and understanding of the effects of ligands on electron-transfer reactions, but only a very small part of the subject has been considered. Before concluding, it seems worth while to place the topics which were discussed into the context of others which have been recognized or are being developed.

The possibility of a doubly bridged, activated complex for electron-transfer reactions was suggested [32] well before proof of any bridging effect for metal ions had been demonstrated. In the experimental investigations on this point, a quite unexpected selectivity has been observed. Thus, in the reactions [33, 34] of either cis-(en)$_2$Co(OH$_2$)$_6^{3+}$ or cis-(H$_2$O)$_4$CrF$_2^+$ with Cr$^{2+}$, though there is an opportunity for double bridging, only the singly bridged activated complex has been observed. However, when cis-(H$_2$O)$_4$Cr(N$_3$)$_2^+$ reacts [35], reaction by the doubly bridged activated complex is much more rapid than by the singly bridged one. The conclusions stated for the three systems referred to are based on direct and unambiguous evidence involving the investigation of the immediate products of the reactions. The conclusions, to be mentioned, based as they are only on the reasonableness of rate-law interpretations, are less certain, but nevertheless seem fairly definite. Each of the following reactions shows a term in the rate law which is first-order in both oxidant and reductant, and inverse first-order in [H$^+$]: the reduction [36] of cis-(NH$_3$)$_5$Co(H$_2$O)OAc$^{2+}$ by Cr$^{2+}$, the isomerization [37] of CrSCN$^{2+}$ catalyzed by Cr$^{2+}$, and chromium exchange between Cr$^{2+}$ and CrH$_2$PO$_2^{2+}$ [38] or CrOAc$^{2+}$ [39]. Taking into account the nature of the processes, it seems likely that they involve simultaneous bridging by the particular anion feature and by a cis hydroxide group. It is to be noted that in all cases involving double bridging, at least one of the partners has a delocalized $\pi$ bond system. The topic is given added interest by the observation [40] that chromium exchange between the fumarato complex of Cr(III) and Cr$^{2+}$ is about 10$^3$ times more rapid than between the acetato complex and Cr$^{2+}$. It is remarkable that the pendant conjugated bond system, though probably not directly involved in providing the connection between Cr$^{2+}$ and the oxidant, can have such a profound effect in the reaction rate.

Halide bridging in reactions proceeding by two-electron changes has been studied to some extent [41, 42, 43]. A systematic comparison between these systems and those involving one-electron changes would be valuable and instructive, particularly if it can be developed to include bridging ligands involving conjugated bond systems. The system Pt(II)—Pt(IV) is at present the only well-defined example of a transition metal two-electron redox system known to react via a bridged activated complex, and some new preparative chemistry would need to be done to provide the comparisons which it seems desirable to make.

The systematic study of the effects in inner-sphere reactions of ligands not directly involved in bridging is now under way [44-48]. The measurement of the isotopic fractionation in ligands cis and trans to the bridging group in the reduction of Co(III) and Cr(II) complexes was referred to in the previous chapter. The effects on the rates of replacing $NH_3$ by $H_2O$, $OH^-$, and other anions have been noted in a number of instances. These effects can be quite large, and the point of interest is to discover the mechanism by which they are exerted. Factors such as the ease of stretching the ligand–oxidizing ion bond and the effect of the ligand in affecting the energy of the acceptor orbital on the metal ion have been recognized as significant. Ligand effects can also be studied for the reducing agent, and a number of systems are known [49, 50] in which $Cr^{2+}$, in attacking a ligand on the oxidant, also incorporates groups present in solution into the coordination sphere in being oxidized to the $3+$ oxidation state. Part of the effect is traceable to the stabilization of the higher oxidation state by the ligand, but it is doubtful that a simple correlation between thermodynamic stability and rate will be applicable to all cases.

Finally, the effect of ligands on the rates of electron transfer by outer-sphere mechanisms needs to be mentioned. Some systematic investigations in this field have been made, but most have been concerned with testing the validity of the Marcus correlation [51]. A re-emphasis on the problem of understanding the rates of self-exchange now seems to be called for. Ligands exert profound effects on the rates, even in systems in which the geometrical factors appear to be very small [52]. Thus, the self-exchange rate for $Ru(en)_3^{2+,3+}$ is of the order of $10^3$ $M^{-1}$ $sec^{-1}$ under conditions for which a self-exchange rate in excess of $10^7$ $M^{-1}$ $sec^{-1}$ for $Ru(o\text{-phen})_3^{2+,3+}$ is indicated. It is clearly of interest to increase the shielding of the $t_{2g}$ electrons by modifying the ligands so as to reduce the rate of self-exchange still further, and then to undertake a systematic study of ligands which, when introduced into the coordination sphere of the complexes, will bring about delocalization of the $t_{2g}$ electrons. Here, as elsewhere in the field, the preparative work, which is often tedious and frustrating, will likely prove to be the step which determines the rate of further progress.

# REFERENCES

1. J. Silverman and R. W. Dodson, *J. Phys. Chem.* **56**, 846 (1952).
2. P. George and J. Griffith, "The Enzymes," Vol. 1, p. 347. Academic Press, New York, 1959.
3. A. M. Zwickel and H. Taube, *Discussions Faraday Soc.* **29**, 42 (1960).
4. J. F. Endicott and H. Taube, *J. Am. Chem. Soc.* **86**, 1686 (1964).
5. J. A. Stritar, Ph.D. Thesis, Stanford University, 1967.
6. A. E. Ogard and H. Taube, *J. Am. Chem. Soc.* **80**, 1084 (1958).
7. D. L. Ball and E. L. King, *J. Am. Chem. Soc.* **80**, 1091 (1958).
8. J. P. Candlin and J. Halpern, *Inorg. Chem.* **4**, 766 (1965).
9. J. P. Candlin, J. Halpern, and D. L. Trimm, *J. Am. Chem. Soc.* **86**, 1019 (1964).
10. J. H. Espenson, *Inorg. Chem.* **4**, 121 (1965).
11. H. Diebler and H. Taube, *Inorg. Chem.* **4**, 1029 (1965).
12. A. Haim, *Inorg. Chem.* **7**, 1475 (1968).
13. H. Taube and H. Myers, *J. Am. Chem. Soc.* **76**, 2103 (1954).
14. The coefficients for the alternatives ($[Co(NH_3)_6^{3+} \cdot X^-]^{2+}$ ($Cr^{2+}$) and ($[Co(NH_3)_6^{3+}]$) ($CrX^+$) which are, in principle, possible, cannot be evaluated because the equilibrium constants corresponding to the associations to form $Co(NH_3)_6^{3+} \cdot X^-$ and $CrX^-$ have not, as yet, been evaluated with a useful degree of accuracy.
15. D. A. Buckingham, I. I. Olson, A. M. Sargeson, and H. Satrapa, *Inorg. Chem.* **6**, 1027 (1967).
17. P. Dodel and H. Taube, *Z. Physik. Chem.* **44**, 92 (1965).
18. $Q_F$ for $Cr^{3+}$ at $\mu = 0.5$ is $\sim 2 \times 10^5$; cf. A. S. Wilson and H. Taube, *J. Am. Chem. Soc.* **74**, 3509 (1952).
19. $Q_{Cl^-}$ at $\mu = 4.4$ is 0.1; cf. H. S. Gates and E. L. King, *J. Am. Chem. Soc.* **80**, 5011 (1958).
20. Approximate values of equilibrium constants are reported by J. F. Endicott and H. Taube, *Inorg. Chem.* **4**, 437 (1965).
21. J. Hudis and A. C. Wahl, *J. Am. Chem. Soc.* **75**, 4153 (1953).
22. H. J. Price and H. Taube, *Inorg. Chem.* **7**, 1 (1968).
23. H. Taube, *Progr. Inorg. Radiochem.* **1** (1959).
24. E. S. Gould and H. Taube, *J. Am. Chem. Soc.* **86**, 1318 (1964).
25. A key observation in this connection was made by E. S. Gould, *J. Am. Chem. Soc.* **87**, 4731 (1965), in showing that $Cr^{2+}$ does not react with terephthalato-pentaamminecobalt(III) significantly more rapidly than with the benzoato complex, contrary to reports made earlier. See [26].
26. H. Taube, *Advan. Chem. Ser.* **49**, 107 (1965).
27. E. M. Kosower and E. J. Poziomek, *J. Am. Chem. Soc.* **85**, 2035 (1963).
28. R. D. Butler and H. Taube, *J. Am. Chem. Soc.* **87**, 5597 (1965).

29. M. Barrett, to be published.
30. F. Nordmeyer and H. Taube, *J. Am. Chem. Soc.* **90**, 1162 (1968).
31. R. Gaunder, work in progress, Stanford University.
32. R. J. Marcus, B. J. Zwolinski, and H. Eyring, *J. Phys. Chem.* **58**, 432 (1954).
33. W. Kruse and H. Taube, *J. Am. Chem. Soc.* **82**, 526 (1960).
34. Y. T. Chia and E. L. King, *Discussions Faraday Soc.* **29**, 109 (1960).
35. R. Snellgrove and E. L. King, *J. Am. Chem. Soc.* **84**, 4609 (1962).
36. K. D. Kopple and R. R. Miller, *Proc. Chem. Soc.* 306 (1962).
37. A. Haim and N. Sutin, *J. Am. Chem. Soc.* **88**, 439 (1966).
38. K. A. Schroeder and J. H. Espenson, *J. Am. Chem. Soc.* **89**, 2548 (1967).
39. E. Deutsch and H. Taube, *Inorg. Chem.* **7**, 1532 (1968).
40. H. Diaz, work in progress, Stanford University.
41. F. Basolo, P. H. Wilks, R. G. Pearson, and R. G. Wilkins, *J. Inorg. Nucl. Chem.* **6**, 161 (1958).
42. F. Basolo, M. L. Morris, and R. G. Pearson, *Discussions Faraday Soc.* **29**, 80 (1960).
43. W. R. Mason and R. C. Johnson, *Inorg. Chem.* **4**, 1258 (1958).
44. P. Benson and A. Haim, *J. Am. Chem. Soc.* **87**, 3826 (1965).
45. R. D. Cannon and J. E. Earley, *J. Am. Chem. Soc.* **88**, 1872 (1966).
46. A. Haim and N. Sutin, *J. Am. Chem. Soc.* **88**, 434 (1966).
47. D. E. Pennington and A. Haim, *Inorg. Chem* **5**, 1887 (1966).
48. C. Bifano and R. G. Linck, *J. Am. Chem. Soc.* **89**, 3945 (1967).
49. H. Taube, *J. Am. Chem. Soc.* **77**, 4481 (1955).
50. J. B. Hunt and J. E. Earley, *J. Am. Chem. Soc.* **82**, 5312 (1960).
51. R. A. Marcus, *J. Phys. Chem.* **67**, 853 (1963); *J. Chem. Phys.* **43**, 679 (1965).
52. T. J. Meyer and H. Taube, *Inorg. Chem.* **7**, 2369 (1968).

# IV

## INDUCED ELECTRON TRANSFER

### I. INTRODUCTION

In the preceding chapter, the possibility of and evidence for the transfer of an electron from an external source (e.g., $Cr^{2+}$ or $Eu^{2+}$) through an extended bond system was considered. The observations made on these systems suggest that low-lying unoccupied orbitals in the bridging ligand play a role in the electron-transfer process. Electron transfer through extended bond systems can also be realized in systems undergoing chemistry quite different from that considered thus far. Electron transfer can be induced by various kinds of operations on the ligand, and one means of inducing internal electron transfer will be dealt with in this chapter.

Consider the complex

$$\left[ (NH_3)_5 CoN \left\langle\!\!\bigcirc\!\!\right\rangle \begin{matrix} H \\ -C-OH \\ H \end{matrix} \right]^{3+}$$

which contains the one-electron oxidizing center Co(III) and a function, the carbinol group, which requires a two-electron oxidation for stable products to be formed. Despite the fact that

73

Co(III) and the carbinol function are contained in the same molecule, the internal redox reaction is extremely slow, both because a cobalt(III) ammine is a rather sluggish oxidant in acting on a two-electron reducing agent, and because, to be oxidized readily by a one-electron oxidant, the carbinol function requires reagents such as Ce(IV), Mn(III) and $Co^{3+}$(aq) having values of $E^0$ in the range of 1.5 V and above. Except under extreme conditions, the disproportionation reaction is not observed. An interesting possibility, however, opens up when a strong one-electron oxidant (to be referred to as the external oxidant) is added to a solution containing **I**. External oxidants even as powerful as $Co^{3+}$(aq) are not able to affect the Co(III) center directly, but they can attack the oxidizable carbinol function to generate organic radicals. If, by the action of the external oxidant, **I** is converted to **II**

$$\left[ (NH_3)_5 CoN \bigcirc - \overset{H}{\underset{\bullet}{C}}OH \right]^{3+}$$

(II)

or to some other species with the associated ligand carrying an odd electron, internal electron transfer may ensue

$$(II) \longrightarrow (NH_3)_5 Co^{II} N \bigcirc - \overset{H}{\underset{\underset{O}{\parallel}}{C}} + H^+$$

(III)

the internal oxidant thereby being reduced to the 2+ state. Now Co(II) is expected to be very labile to substitution (it is, in fact, possible that the Co—N bond is severed in the very act of reduction of the cobalt center) so that in acidic solution, **III** will dissociate to $Co^{2+}$, $NH_4^+$, and

$$\left[ HN \bigcirc - \overset{H}{\underset{\underset{O}{\diagdown}}{C}} \right]^{+}$$

The net result, then, is that one equivalent of the external oxidant and one of the internal oxidant are consumed to bring about the two-electron oxidation of the oxidizable group on the ligand. The expectation which has been suggested for systems of this kind is realized. The systems in which electron transfer can be induced by the means described have several features of interest. One of these is the possibility which is opened of studying electron transfer from a radical located at a remote position through an intervening bond system to the acceptor center, as a function of the structure of the ligand. When electron transfer is induced by the means outlined, the excess electron is produced in the σ bond system. There is no assurance that the rules governing the facility of electron transfer will be identical with those applicable to reducing agents that act without causing irreversible change in the ligands which mediate in electron transfer. If—and this remains to be shown—the action of the external oxidant takes place independently of the internal electron transfer act, i.e., if the internal oxidant does not influence the action of the external one, the possibility opened up for internal electron transfer is a powerful tool for probing the mechanism of oxidation by the external oxidant. Specifically, we have a means of classifying the mode of action of the external oxidant according to whether a one-electron change takes place [to induce reduction of the Co(III) center] or whether a two-electron change takes place (so that the radical state is avoided and $Co^{2+}$ is not formed).

Work with systems of this kind has not progressed very far, but enough has been learned so that a great potential for this approach is apparent. The account which follows begins with a brief history of the subject, then deals with the important question of whether the external and internal oxidants act in concert or independently, and ends with a survey of some of the applications which have been made of the induced electron-transfer effect.

Cobalt complexes are emphasized because the effect so far has been demonstrated only for them. The reasons underlying the unique position of cobalt are worthy of consideration. For induced electron transfer to be observed directly, that is, by showing that the reduced form of the internal oxidant is among the products of reaction, the product of internal electron transfer must be substitution-labile. If intermediate **III** persists long enough so that the external oxidant reoxidizes Co(II) to Co(III), only

$$\left[(NH_3)_5CoN \bigcirc\hspace{-1.2em}\bigcirc - C\!\!\begin{array}{c}H\\ \\O\end{array}\right]^{3+}$$

would be observed as a product, and it would then be difficult to show that interesting chemistry had led to the formation of this product. Just this situation obtains when

$$\left[(NH_3)_5RuN \bigcirc\hspace{-1.2em}\bigcirc - \begin{array}{c}H\\COH\\H\end{array}\right]^{3+}$$

is oxidized. Only

$$\left[(NH_3)_5RuN \bigcirc\hspace{-1.2em}\bigcirc - C\!\!\begin{array}{c}H\\ \\O\end{array}\right]^{3+}$$

is formed, irrespective of whether the external oxidant acts by a one-or by a two-electron change. Ruthenium(II) is much less labile than cobalt(II), and the ruthenium(II) complex resulting from internal electron transfer is reoxidized by the external oxidant before the heteroligand is lost from it.

## II. History

Reports of the formation of $Co^{2+}$ in the manipulation of cobaltammine complexes appear in the classic literature of coordination chemistry. A specific system in which mention of this kind of effect [1] is made involves the action of $H_2O_2$ on $(NH_3)_5CoNCS^{2+}$. This system has more recently been subjected to a careful reexamination [2, 3]. The first study in which the significance of the effect was explored was made with $(NH_3)_5CoC_2O_4H^{2+}$ as the substrate [4]. The work done with this complex showed that one-electron oxidants such as Ce(IV) and $Co^{3+}$(aq) bring about reduction of the internal oxidant Co(III), while the external oxidants $Cl_2$(aq) and $H_2O_2$ [the latter in the presence of Mo(VI) as catalyst], which can reasonably be considered to act by two-electron changes, produce $(NH_3)_5CoOH_2^{3+}$. In both sets of reaction, the oxalate ligand is converted to $CO_2$ ; in the former set the net two-electron change is shared between the internal and external oxidants, in the latter it is brought about by the external oxidants. Efforts made to intercept the intermediate IV,

$$(NH_3)_5Co^{(III)} \cdot [C_2O_4^-]$$
IV

which may be presumed to be formed by one-electron external oxidants before electron transfer takes place, were unsuccessful, and thus little progress was made on the important question of whether the external and internal oxidants act in concert or independently. The latter conclusion was indicated by the observation that the external oxidants act very slowly, and thus without showing evidence for cooperation by the internal oxidant, but qualitative evidence such as this is hardly a sufficient basis for a firm conclusion. Analogous chemistry has been reported for the oxidation of

$$\left[ (NH_3)_5Co-O-\overset{\overset{\displaystyle O}{\|}}{C}-\bigcirc-\overset{\overset{\displaystyle H}{}}{C}\underset{\displaystyle O}{\diagdown} \right]^{2+}$$

but only a cursory investigation of this system was made [5]. The work by Candlin and Halpern [6] provided the first evidence in systems of this kind that an intermediate is formed, and thus that internal electron transfer takes place after the external oxidant has acted. These results and others bearing on this point will be discussed more fully in the next section.

### III. Evidence for the Formation of an Intermediate

The rate law for the reaction of $(NH_3)_5CoO_2CH^{2+}$ with $MnO_4^-$ has the form [6]

$$-d[(NH_3)_5CoO_2CH^{2+}]/dt = k[(NH_3)_5CoO_2CH^{2+}][MnO_4^-] \quad \text{(IV-1)}$$

Though the specific rate defined as above does not depend on the concentration of $MnO_4^-$, the chemistry does. Both $Co^{2+}$ and $(NH_3)_5CoOH_2^{3+}$ are products of the reaction, and the ratio in which they are formed depends on $[MnO_4^-]$. The function

$$[Co(III)]/[Co(II)] = 3.0 \times 10^2[MnO_4^-] \quad \text{(IV-2)}$$

was offered by Candlin and Halpern as describing the results on the product ratio at $0.1°C$. The observations cited prove that the reaction mechanism is of the form

$$(NH_3)_5CoO_2CH^{2+} + MnO_4^- \rightarrow Int \quad \text{(IV-3)}$$

$$Int \rightarrow Co^{2+} \quad \text{(IV-4)}$$

$$Int + MnO_4^- \rightarrow (NH_3)_5CoOH_2^{3+} \quad \text{(IV-5)}$$

but of course do not of themselves settle the question of the identity of the intermediate which must be postulated. For the present purposes, the important conclusion is that an intermediate is generated by the action of the external oxidant on the substrate complex, and that only after the intermediate is formed is the choice made between forming $Co^{2+}$ or $(NH_3)_5CoOH_2^{3+}$ as products. That is, the external oxidant and internal oxidant do not cooperate in the process which leads to production of $Co^{2+}$ and the net two-electron oxidation of the ligand.

The conclusion just stated applies also to a number of other systems which have been investigated in detail, and it is important to one of the applications of the effect which will be made. This application depends on the fact that when the external oxidant acts by a two-electron change, a Co(III) complex is produced, but if it acts by a one-electron change, both $Co^{2+}$ and Co(III) can appear among the products. But for any conclu-

TABLE IV-1

THE RATE OF OXIDATION BY Ce(IV) OF LIGANDS
AS A FUNCTION OF ASSOCIATED LEWIS ACID[a]

| Ligand | On $H^+$ | On $(NH_3)_5Co^{3+}$ |
|---|---|---|
| ![structure: pyridine ring with C(H)(H)—OH substituent] | $8.3 \times 10^{-3}$ | $9.2 \times 10^{-3}$ |
| ![structure: pyridine ring with CH(H)—OH substituent] | $5.1 \times 10^{-3}$ | $5.5 \times 10^{-3}$ |
| ![structure: pyridine ring with C(=O)H substituent] | $7.8 \times 10^{-4}$ | — |
| ![structure: pyridine ring with C(=O)H substituent] | $2.0 \times 10^{-3}$ | — |

[a] In 1 $M$ $H^+$ at 25° and $\mu = 2.0$.

sion about mechanism to be applicable when the ligand is not coordinated to Co(III), we must be certain that Co(III) does not promote one path relative to another when the external oxidant acts. Observations on the rate of oxidation of ligands support the conclusion reached in the preceding paragraph. If Co(III) assists in the oxidation of the ligand by the external agents, this should be reflected in an increased rate of reaction for the ligand complexed by Co(III) compared to the rate when the ligand is free. Allowance must, of course, be made for the fact that the charge on the cobalt atom will affect the rate of reaction, but when the oxidizable group is several bonds removed from $Co^{3+}$, this effect is expected to be small.

Some data [7] on the rate of oxidation of isomers of pyridine-methanol to the corresponding aldehydes when coordinated to the pentaamminecobalt(III) radical and when coordinated to $H^+$ are shown in Table IV-1.

The rate differences recorded in Table IV-1 are small, and do not speak for any substantial assistance by Co(III) in the attack on the ligand by the external oxidant.

## IV. The Oxidation of Pentaamminepyridinemethanolcobalt(III) by Cerium(IV)

The induced effect has been most thoroughly and extensively investigated for species of the type

$$\left[ (NH_3)_5CoN \underset{CH_2OH}{\bigodot} \right]^{3+}$$

and the results for this system [7] will be outlined in some detail so as to illustrate the use of the induced effect as a probe of mechanism, and to indicate the kind of problems which arise in interpreting some of the detailed information which is yielded by the probe.

When

$$\left[(NH_3)_5CoN \bigcirc \overset{H}{\underset{H}{-C}OH}\right]^{3+}$$

is oxidized by Ce(IV), Co(III), Ag(II), or $S_2O_8^{2-}$ (the latter as catalyzed by Ag$^+$), Co$^{2+}$ is among the products, and we conclude that each of these reagents attacks, at least in part, by one-electron changes. (The detailed results which establish the extent to which the primary attack is by a one-electron change will be outlined presently.) However, when Cl$_2$, H$_2$O$_2$ catalyzed by Mo(VI), or $Cr_2O_7^{2-}$ in acid solution act on the complex, only

$$\left[(NH_3)_5CoN \bigcirc -\overset{H}{C}\diagdown_O\right]^{3+}$$

is formed as the product of oxidation, at least if the secondary oxidation of the product pyridinecarboxaldehyde complex is discounted. These results indicate that the last group of oxidants attack the alcohol function without forming a radical intermediate, and thus by a two-electron change. This conclusion is particularly reassuring for $Cr_2O_7^{2-}$ because, in this case, other evidence has already established it with a considerable degree of confidence [7, 8]. The result obtained should be regarded as testing the induced effect as a probe of mechanism, rather than as contributing materially to the strength of the conclusion on the mechanism by which $Cr_2O_7^{2-}$ oxidizes the carbinol function. To complete the test of mechanism by the internal electron-transfer probe, further experiments are necessary, because the possibility must be covered that Cr(VI) may actually produce a radical intermediate

$$\left[(NH_3)_5CoN \bigcirc -\overset{H}{\underset{\bullet}{C}OH}\right]^{3+}$$

but that it then so rapidly oxidizes **V** further, forming $[(NH_3)_5CoNC_5H_4CHO]^{3+}$, that there is no opportunity for the internal electron transfer which would lead to the formation of $Co^{2+}$. This possibility has been eliminated by using the oxidizing mixture $Co^{3+}(aq) + Cr_2O_7^{2-}$. The former reagent is known to produce a radical from the carbinol function, and under appropriate conditions attacks this function more rapidly than does $Cr_2O_7^{2-}$. When this oxidizing mixture is used, $Co^{2+}$ still appears as a product of the reaction, and thus we have proof that $Cr_2O_7^{2-}$ does not interfere so completely with the electron-transfer act that the production of $Co^{2+}$ is thereby prohibited.

The mechanism of attack of an oxidant can be dramatically altered by the use of catalysts, and the induced transfer probe can be a powerful tool [6] in investigating catalytic effects in redox reactions. When Ce(IV) acts in the presence of $Ag^+$, the rate of reaction becomes independent of the substrate concentration, but the mode of action of the oxidizing system still is that of a one-electron agent. Ce(IV) rapidly oxidizes $Cr(H_2O)_6^{3+}$ to Cr(VI) in acidic perchlorate media. When Ce(IV) is slowly added to Cr(III) in the presence of the 4-pyridinemethanol complex, a major fraction of the oxidizing power is consumed by the induced oxidation of the complex to the corresponding carboxyaldehyde, with no formation of $Co^{2+}$. This rapid, induced oxidation must be produced by the intermediate (IV) or (V) oxidation states of chromium. Although more needs to be done to understand this reaction, it seems probable that Cr(V) is the major oxidant which, as has been concluded [8, 9] from studies with $Cr_2O_7^{2-}$ as the oxidant, rapidly oxidizes the alcohol by a two-electron change.

The quantitative results on product ratio as a function of the concentration of the external oxidant bring a number of puzzling effects to light. Fig. IV-1 shows the product ratio (here defined as $[Co^{2+}]/[(NH_3)_5CoNC_5H_4CHO]^{3+}$ for the uncatalyzed Ce(IV) oxidation of the 4-pyridinemethanol complex as a function of Ce(IV) concentration. As the Ce(IV) concentration increases, this ratio decreases as would be expected on the basis of the

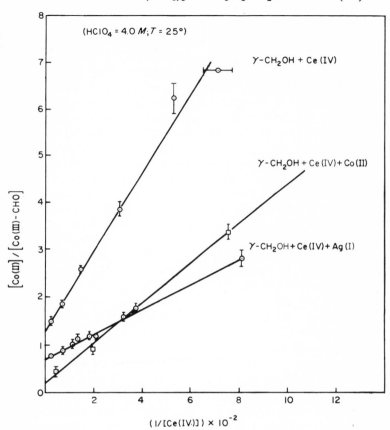

FIG. IV-1. Difference between stoichiometry of the reaction of pentaammine 4-pyridinemethanolcobalt(III) with different oxidants.

mechanism type presented earlier [equations (IV-3), (IV-4), and (IV-5)], but the ratio does not decrease to zero, as would be expected on the basis of this mechanism, when [Ce(IV)] increases indefinitely. These results show that two intermediates are formed by the action of the external oxidant. Both can undergo internal electron transfer—thus, we conclude that Ce(IV) acts solely as a one-electron oxidant—but one intermediate, the contribution of which to the stoichiometry is measured by the intercept at $1/[Ce(IV)] = 0$, undergoes internal electron transfer with such facility that this act is not interrupted by the external oxidant, while the other, the behavior of which is defined by the slope of the line in Fig. IV-1, lives sufficiently long so that further reaction with external oxidant competes with internal electron transfer.

An immediate question which arises from the conclusion that two intermediates result from the oxidation of the carbinol function by a single one-electron external oxidizing agent is whether the intermediates are formed by parallel processes or result from a single act performed on the oxidizable function by the external oxidant. According to the latter alternative, the ratio in which the two intermediates are formed should be independent of the nature of the external oxidant. Figure IV-1 also presents data obtained for the system $Ce(IV) + Ag^+$, where $Ce(IV)$ but not $Ag^+$ is consumed in the net change, and the latter reagent acts as catalyst. In this system, some higher oxidation state of silver, probably Ag(II), is the agent which attacks the carbinol, and the partition between the trappable and nontrappable intermediate as defined by the intercept in the relevant plot in Fig. IV-1 at $1/[Ce(IV)] = 0$ is seen to be different from that obtained with Ce(IV) alone. Thus, the conclusion is indicated that the two different intermediates are formed by parallel processes in the attack of the substrate by the external oxidant. This conclusion is contrary [10] to that reached in a study of the product ratio for

$$\left[ (NH_3)_5 CoO_2C - \underset{\bigcirc}{\bigcirc} - CH_2OH \right]^{2+}$$

as it is acted upon by external oxidants. Here, too, two intermediates are formed, one readily trappable by external oxidant and
the other nontrappable, but here the partition is almost the same
whether the external oxidant is Ce(IV) in perchlorate, Co(III) in
perchlorate, or Co(III) in sulfate. An approximate correspondence
can always be attributed to coincidence, thus avoiding the neces-
sity of attaching deeper meaning to it, but the probability of its
being merely coincidental in the present instance is reduced by
the observation [7] that, in the oxidation of pentaammine-3-
pyridinemethanolcobalt(III), the partition between the two
intermediates is also very nearly the same whether Ce(IV),
Ag(II), or Co(III) is the oxidant. In the absence of a deeper
understanding of the problem, which might reveal why the
partition is (almost) independent of the nature of the oxidant in
two of the systems but clearly dependent on it in the other, the
position will be taken that the two intermediates are formed by
parallel rather than interrelated events. The basis for this
decision is that the evidence for parallel events in the 4-pyridine-
methanol case is clear, and the agreement of the partition ratios
in the others is not so close as to make it impossible that sheer
coincidence is responsible.

If this decision is accepted as a conclusion the following
mechanism accounts for the results:

$$Ce(IV) + (NH_3)_5CoNC_5H_4CH_2OH^{3+} \rightarrow \underset{VI}{Int} + Ce^{3+} \qquad (IV\text{-}6)$$

$$VI \rightarrow Co^{2+} + NH_4^+ + HNC_5H_4CHO^+ \qquad (IV\text{-}7)$$

$$VI + Ce(IV) \rightarrow (NH_3)_5CoNC_5H_4CHO^{3+} + Ce^{3+} \qquad (IV\text{-}8)$$

$$Ce(IV) + (NH_3)_5CoNC_5H_4CH_2OH^{3+} \rightarrow \underset{VII}{Int} + Ce^{3+} \qquad (IV\text{-}9)$$

$$VII \rightarrow Co^{2+} + NH_4^+ + HNC_5H_4CHO^+ \qquad (IV\text{-}10)$$

$$VII + Ce(IV) \rightarrow (NH_3)_5CoNC_5H_4CHO^{3+} \qquad (IV\text{-}11)$$

On applying the steady state approximation to the intermediates
**VI** and **VII**, and assuming that $k_{11}^*/k_{10}$ is zero, it can be shown
that the product ratio $[Co^{2+}]/[Co(III)]$ is given by the function

$$\frac{k_7}{k_8}\left[\frac{k_9}{k_6} + 1\right]\frac{1}{[Ce(IV)]} + \frac{k_9}{k_6} \qquad \text{(IV-12)}$$

The intercepts in Fig. IV-1 measure $k_9/k_6$, which is the ratio in which the nontrappable and trappable intermediates are formed. From this ratio and the slope of the lines in Fig. IV-1, $k_7/k_8$, which is the specific rate of reaction of the intermediate **VI** by internal electron transfer compared to that of reaction with external oxidant, can be calculated.

The ratio $k_7/k_8$ is important in defining an intrinsic property of the intermediate, and also in providing an opportunity for measuring the absolute rate of electron transfer from a radical located at a position remote from the Co(III) acceptor center. The data in Fig. IV-1 provide an illustration of the use of the reactivity ratio in the former sense. According to function (IV-12), if $k_7/k_8$ is constant, the ratios of the slopes in Fig. IV-1 can be calculated from the intercepts shown.

It should be noted that in the Ag-catalyzed system, the concentration of the intermediate, presumed to be Ag(II), which attacks the complex is kept low compared to that of Ce(IV) [the reaction of Ag(II) with an alcohol appears to be very rapid], and thus Ce(IV) rather than Ag(II) is the species which attacks the trappable radical intermediate. When the calculations are performed, the ratios for the uncatalyzed and Co(II)-catalyzed reactions agree, as required if the properties of the trappable intermediate are independent of the system, but disagree with that calculated for the Ag⁺-catalyzed reaction. The disagreement in the expected and calculated values of $k_7/k_8$ is only approximately a factor of 2, but it is large enough to be outside experimental error, and it therefore indicates that the properties of the intermediate can be modified by the environment. The cause of the change in properties with change in medium is not known, but a likely possibility is that the radical in question to some extent forms a complex with Ag⁺. Not all variations in the conditions produce a change in $k_7/k_8$, and with the 3-pyridinemethanol complex the difference in environment is not reflected in a noticeably changed reactivity pattern.

To measure the rate of internal electron transfer, the absolute value of $k_8$ must be known. Such measurements have not yet been made for radicals derived from the carbinols under investigation. Oxygen is known to react rapidly with carbon radicals, and has is found to be effective in inhibiting the formation of $Co^{2+}$. In the 4-pyridinemethanol systems, it interferes with internal electron transfer only for the more trappable intermediate, and for this intermediate, with Ce(IV) in $4\,M$ $HClO_4$ as trapping agent, $k_7/k_8$ is about $4 \times 10^{-4}\,M$. Even when $k_8$ is assigned the maximum value of $10^{10}\,M^{-1}\,sec^{-1}$ expected for a diffusion-controlled reaction, $k_7$ is calculated to be $4 \times 10^6$, and thus a lifetime for the trappable intermediate of at least $10^{-7}\,sec$ is indicated. For the system with

$$\left[ O_2C - \left\langle \bigcirc \right\rangle - CH_2OH \right]^-$$

as ligand, a roughly similar lifetime for the corresponding radical was calculated.

Even when the actual value of $k_7$ cannot be calculated, a comparison of the ratio $k_7/k_8$ for a common radical in different situations is of interest. It is reasonable to suppose that the rate of reaction of a radical such as

$$\overset{\displaystyle H}{\underset{\displaystyle \cdot}{R-C-OH}}$$

with an external oxidant will be rather insensitive to the nature of R, at least when the external oxidant is quite reactive. Thus, the values of the ratio $k_7/k_8$ can provide a reasonably good measure of the relative rates of internal electron transfer. Data of this kind have been collected for several systems, and the results are summarized in Table IV-2.

## TABLE IV-2

RELATIVE RATES OF INTERNAL ELECTRON TRANSFER
THROUGH VARIOUS BOND SYSTEMS[a]

| Ligand | [$HClO_4$] | $k_9/k_6$ | $k_{10}/k_{11}$ | $k_7/k_8$ |
|---|---|---|---|---|
| 4-Pyridinemethanol | 4.0 | 1.30 | >1 | $3.8 \times 10^{-3}$ |
| 4-Pyridine-$D_2$-methanol | 4.0 | 0.70 | >1 | $3.8 \times 10^{-3}$ |
| 4-Pyridinemethanol | 1.0 | 1.8 | >1 | $(1.5 \pm 1) \times 10^{-2}$ |
| 4-Hydroxymethylbenzoate | 1.0 | 0.8 | >1 | $1.0 \times 10^{-3}$ |
| 3-Pyridinemethanol | 1.0 | 3.0 | $8 \times 10^{-2}$ | $<1 \times 10^{-4}$ |

[a] External oxidant Ce(IV), $k_8$ reactant Ce(IV), $\mu = $ [$HClO_4$].

For each of the systems described, two radicals appear to be formed, which differ with respect to the rate of internal electron transfer ($k_7$ or $k_{10}$) compared to the rate of reaction with external oxidant ($k_8$ or $k_{11}$). Internal transfer is slower in the 3-pyridine-methanol than in the 4-pyridine isomer, in agreement with general expectations. It is interesting also that transfer to Co(III) through the pyridine nitrogen is not very much faster than through a carboxyl group. This result is particularly significant in the context of the fact that the complex

$$(NH_3)_5CoN \langle\bigcirc\rangle - CO_2H^{3+}$$

is reduced [11] by $Cr^{2+}$ much more rapidly than is [12]

$$(NH_3)_5CoO_2C - \langle\bigcirc\rangle - CO_2H^{2+}$$

The rate ratio is approximately 100, but it must be remembered that the dominant path for the reduction of the latter complex probably involves attack by the reducing agent at the adjacent carboxyl rather than at the one corresponding to the reduction of

the pyridine complex, and thus the rate ratio for remote attack
may be as high as 1000. The effect on the rate of reduction of
replacing a carboxyl group as the function leading the electron
to Co by a pyridine N probably has less to do with any local
properties of the group than with the fact that N as part of the
aromatic ring system lowers the energy of an unoccupied orbital;
i.e., the greater rate of reaction for the pyridine complex is related
to the greater ease of populating a low-lying orbital by an electron
derived from $Cr^{2+}$.

Comparisons of the type illustrated in Table IV-2 are of
interest for other kinds of bond systems extending from the
carbinol function to the acceptor center, but they also need to
be made with oxidizable functions other than the carbinol. Some
preliminary experiments with the complex of 4-pyridinecarbox-
aldehyde and with Ce(IV) as oxidant suggest that more production of $Co^{2+}$ takes place than is the case with the 4-pyridine-
methanol complex. Considerations advanced earlier show that
at least one of the intermediates obtained in the one-electron
oxidation of the 4-pyridinemethanol complex undergoes internal
electron transfer rather slowly. This slowness can probably
not be attributed entirely to the requirements for excitation
of the Co center before electron transfer can take place, and
changes in configuration at the carbon radical center must
probably also take place before the electron is passed on to the
acceptor.

Closely related to the issue just raised is the question of the
nature of the two intermediates which are generated when
Ce(IV), $Co^{3+}$(aq), or Ag(II) oxidizes an alcohol. Experiments
with $-CD_2OH$ rather than $-CH_2OH$ as the group being
oxidized showed that the rate of formation of the less trappable
intermediate declines by a factor of about 3.7 when D replaces
H, while for the other intermediate the factor is about 2.0. It
seems possible that the former intermediate arises by hydrogen
atom abstraction (as indicated earlier for intermediate **IV**), and
the latter by electron abstraction from the oxidizable function

$$\begin{matrix} \text{H} \\ -\text{COH} \\ \text{H} \end{matrix} \rightarrow \begin{matrix} \text{H} \\ -\overset{.}{\text{C}}\text{OH} \\ \textbf{IV} \end{matrix} + \text{H}^+ + e \qquad \text{(IV-13)}$$

$$\begin{matrix} \text{H} \\ -\text{COH} \\ \text{H} \end{matrix} \rightarrow \begin{matrix} \text{H} \\ -\text{COH}^+ \\ \text{H} \\ \textbf{VIII} \end{matrix} + e \qquad \text{(IV-14)}$$

Internal electron transfer would be expected to be slow for (IV-14) because the electron-transfer act would demand proton loss from the radical ion. The oxidation of intermediate **VIII** by external oxidants could, however, take place rather readily by hydrogen-atom transfer.

## V. LOW-SPIN Co(II) AS AN INTERMEDIATE

The possibility that a low-spin form of Co(II) is generated as an intermediate in the reduction of a cobaltammine by an inner-sphere mechanism was mentioned in an earlier chapter. Chemical evidence for such an intermediate can, in principle, be obtained by exploiting the reoxidation. If an agent is chosen which can transfer a characteristic group to Co(II) in the act of oxidizing it, then an amminecobalt(III) species bearing the group should appear in the system

$$\text{Co(NH}_3)_5^{2+} + \text{YO} \rightarrow \text{Co(NH}_3)_5\text{Y}^{2+} + \text{O} \qquad \text{(IV-15)}$$

This method is not readily applicable to systems in which a reducing agent such as $\text{Cr}^{2+}$ attacks a cobaltammine complex, because any oxidizing agent reactive enough to oxidize the presumed intermediate readily would also oxidize the reducing agent. However, in the systems under present discussion, the agent attacking the cobaltammine complex is itself an oxidizing agent and can, in principle, reoxidize $\text{Co(NH}_3)_5^{2+}$ if it is produced.

It is unlikely that a reagent such as $\text{Co}^{3+}(\text{aq})$ would be reactive enough to reoxidize $(\text{NH}_3)_5\text{Co}^{2+}$ before the electronically excited species meets another fate, but an anionic species is expected to react more rapidly with it. On this basis, an experiment was done using the oxidizing mixture $Co^{3+}(aq) + Cr_2O_7^{2-}$

acting on

$$\left[ (NH_3)_5CoN \left\langle \bigcirc \right\rangle - CH_2OH \right]^{3+}$$

where $Co^{3+}(aq)$, as an external one-electron oxidant, is a potential source of $(NH_3)_5Co^{2+}$:

$$(NH_3)_5CoNC_5H_4CH_2OH^{3+} + Co^{3+}(aq)$$

$$\rightarrow (NH_3)_5Co^{2+} + HNC_5H_4CHO^+ + Co^{2+}(aq)$$

and $Cr_2O_7^{2-}$ is chosen as being reasonably compatible with the substrate complex and in the hope that it would prove to be very reactive toward $(NH_3)_5Co^{2+}$. To the extent that $(NH_3)_5Co^{2+}$ is formed in the mixture and is reoxidized by $Cr_2O_7^{2-}$, $(NH_3)_5CoOH_2^{3+}$ should appear among the products [the chromium complex formed as the immediate product of oxidizing $(NH_3)_5Co^{2+}$ would aquate rapidly]. The test of the mechanism indicated by the chemistry outlined is easy to apply because $(NH_3)_5CoOH_2^{3+}$ can be separated using ion-exchange techniques from other Co(III) complexes in the system. It has been applied to the system under discussion, but no trace of $(NH_3)_5CoOH_2^{3+}$ was found among the products of reaction. The experiment does not prove that $(NH_3)_5Co^{2+}$ is not formed as the result of internal electron transfer, but it does prove that this intermediate, if produced, undergoes electronic rearrangement to the stable form before it is oxidized by $Cr_2O_7^{2-}$.

Permanganate ion can be expected to be more reactive toward

$(NH_3)_5Co^{2+}$ than is $Cr_2O_7^{2-}$, and systems involving it provide perhaps the best hope of capturing the intermediate by reoxidation. Candlin and Halpern interpreted their observations on the reaction between $(NH_3)_5CoO_2CH^{2+}$ and $MnO_4^-$ by assuming $[(NH_3)_5CoO_2C\cdot]^{2+}$ to be the intermediate which undergoes either internal electron transfer (forming $Co^{2+} + CO_2$) or oxidation by $MnO_4^-$ [forming $(NH_3)_5CoOH_2^{3+} + CO_2$]. However, their results can be equally well understood on the basis that the intermediate in question is low-spin Co(II), of formula $(NH_3)_5Co^{2+}$. New experiments which, in scope, go beyond those of Candlin and Halpern, in fact somewhat favor a mechanism with $(NH_3)_5Co^{2+}$ rather than $[(NH_3)_5CoO_2C\cdot]^{2+}$ as the intermediate.

Most of the new observations [13] were made on solutions in acetate buffers at pH 5.2, rather than at 0.1 $M$ H$^+$, as in the earlier studies. At the higher pH, experiments tracing the fate of permanganate oxygen are feasible, and these have contributed what is perhaps the most significant new information. It is found that permanganate oxygen is incorporated into the product aquo ion. The extent of incorporation increases as the concentration of permanganate increases, reaching a limit of about one-third of the total aquo ion when $1/[MnO_4^-]$ approaches zero. As observed by Candlin and Halpern for solutions at higher acidity, the product ratio $[Co(III)]/[Co^{2+}]$ changes linearly with $[MnO_4^-]$, but in contrast to the conclusion reached by them, we find that this ratio does not approach zero when the concentration of permanganate approaches zero. In the limit of $[MnO_4^-] = 0$, only one-third of the cobalt appears as $Co^{2+}$.

These observations and other tracer results which will not be described here are readily explained by the mechanism

$$(NH_3)_5CoO_2CH^{2+} + MnO_4^- \rightarrow (NH_3)_5Co-O-\underset{\underset{O}{\parallel}}{C}-OMnO + H^+$$

$$\tag{IV-16}$$

$$(NH_3)_5Co-O-\underset{\underset{\displaystyle O}{\|}}{C}-OMnO \rightarrow (NH_3)_5Co^{2+} + MnO_4^{2-} + CO_2 \tag{IV-17}$$

$$(NH_3)_5Co^{2+} + MnO_4^- \rightarrow (NH_3)_5CoOH_2^{3+} + MnO_4^{2-} \tag{IV-18}$$

$$(NH_3)_5Co^{2+} \rightarrow Co^{2+} + NH_4^+ \tag{IV-19}$$

$$(NH_3)_5Co-O-\underset{\underset{\displaystyle O}{\|}}{C}-OMnO \rightarrow (NH_3)_5CoOH_2^{3+} + CO_2 + MnO_4^{3-} \tag{IV-20}$$

The reaction of $MnO_4^-$ with the formate ion can be assumed to be a two-electron change, so that Co, C, and Mn can be assigned oxidation states of III, IV, and V, respectively, in the intermediate $(NH_3)_5CoO_2COMnO_3$. In reaction (IV-17) Mn(V) is assumed to reduce Co(III) to $(NH_3)_5Co^{2+}$. The low-spin Co(II) can be reoxidized by $MnO_4^-$ [reaction (IV-18)] to place permanganate oxygen on Co, or be deactivated to Co(II) of normal electronic structure. To explain the finite value of the product ratio $[Co(III)]/[Co^{2+}]$ as $[MnO_4^-]$ approaches zero, reaction (IV-20) is introduced: the intermediate formed in reaction (IV-16) in part hydrolyzes to Co(III) and Mn(V), thereby preserving the oxidation state of Co(III). If two-thirds of the reaction is assumed to proceed this way, it is clear that the maximum labeling by permanganate oxygen of the aquo product will approach one-third when $[MnO_4^-]$ increases without limit.

Though the mechanism proposed provides a satisfactory explanation of the facts, it is by no means unique. The Candlin and Halpern mechanism can be adapted so that it too will explain the general mechanistic features. In their mechanism, a one-electron oxidation of $(NH_3)_5CoO_2CH^{2+}$ by $MnO_4^-$ is assumed, producing

$$\left[ (NH_3)_5Co-O-\underset{\underset{\displaystyle O}{\|}}{C^{\cdot}} \right]^{2+}$$

this intermediate can undergo two fates:

$$\left[ (NH_3)_5Co-O-\underset{\underset{O}{\|}}{C}\cdot \right]^{2+} + MnO_4^-$$

$$\rightarrow (NH_3)_5CoOH_2^{3+} + CO_2 + MnO_4^{2-} \qquad (IV-21)$$

$$\left[ (NH_3)_5Co-O-\underset{\underset{O}{\|}}{C}\cdot \right]^{2+} \rightarrow Co^{2+} + CO_2 \qquad (IV-22)$$

To explain the finite value of $[Co(III)]/[Co^{2+}]$ when $[MnO_4^-]$ approaches zero, a process involving a net two-electron oxidation of the complex needs to be introduced:

$$(NH_3)_5CoO_2CH^{2+} + MnO_4^- \rightarrow (NH_3)_5CoOH_2^{3+} + CO_2 + MnO_4^{3-}$$

This mechanism makes it more difficult to understand the efficient transfer of oxygen from $MnO_4^-$ to cobalt, and is also less satisfactory than the other in explaining other features of the labeling results. The weaknesses are not so severe, however, that this mechanism needs to be discarded in favor of the other, and the formation of low-spin $(NH_3)_5Co^{2+}$ as an intermediate remains an interesting possibility, but has not been proved. The extension of the studies to other induced oxidation reactions may, however, lead to definite conclusions on this point of interest.

## VI. Other Systems

One of the mechanisms suggested for the reaction of $MnO_4^-$ with $(NH_3)_5CoO_2CH^{2+}$ features the formation of an intermediate by a two-electron oxidation of the ligand, this intermediate then having the capability of reducing the Co(III) center by a one-electron change. Just this mechanistic feature has been introduced [2] to explain the reduction of Co(III) which accompanies the oxidation by $H_2O_2$ of $(NH_3)_5CoNCS^{2+}$. The rate law (except for the dependence on concentration of acid) for the reaction is simple, the rate of consumption of the isothiocyanate complex being found to be first-order in $H_2O_2$ and in $(NH_3)_5CoNCS^{2+}$. Both $Co^{2+}$ and $Co(NH_3)_6^{3+}$ [3] are observed as cobalt-containing products. In contrast to the systems thus far discussed, the ratio $[Co^{2+}]/[Co(III)]$ in the products does not depend on the concentration of the external oxidant, but it does vary linearly with $[H^+]$. An interesting property of an intermediate state in the oxidation of $NCS^-$ by $H_2O_2$ is revealed by the study of induced electron transfer, but the oxidation is itself so complex, involving as it does an overall eight-electron change, that the intermediate state in question cannot be identified from the evidence.

Other methods besides oxidation can be imagined for damaging a ligand so that electron transfer to Co(III) can take place. When reducing agents are used, the ligand can be saved from net reduction by being associated with Co(III). This is the case for the reaction of $(NH_3)_5CoO_2CCHO^{2+}$ or of $(NH_3)_5CoNO_3^{2+}$ with $Cr^{2+}$. The free ligands are readily reduced by $Cr^{2+}$, but when they are associated with Co(III), this center absorbs the reducing electron and the ligand transfers without suffering reduction. Other cases have been encountered in which the external reducing agent may preferentially act on a reducible group on the ligand. The reduction [14] of the isomeric forms of the pentaammine complexes of nitrobenzoic acid provides some interesting examples of this kind of behavior. When $Cr^{2+}$ is added to a solution of the *p*-nitrobenzoatopentaamminecobalt(III) in 1.2 $M$ $H^+$, the reducing agent is consumed rapidly, but reduction of the Co(III) center does not ensue until more than 1 equivalent of $Cr^{2+}$ has been added. With the *meta* isomer as ligand, approximately 4 equivalents of $Cr^{2+}$ are needed before the Co(III) center is reduced; with the *ortho* derivative, reduction of the Co(III) sets in with the first addition of $Cr^{2+}$, and the reducing agent partitions between reducing Co(III) and the ligand. With the *meta* derivative, apparently the stage

$$\left[ (NH_3)_5CoO_2C \underset{}{\bigcirc} N{=}N \underset{}{\bigcirc} CO_2Co(NH_3)_5 \right]^{4+}$$

is reached before reduction of Co(III) can compete with reduction of the ligand. Electron transfer between *para* substituents is expected to be more facile than when the substituents are in *meta* positions; in accord with this expectation internal transfer to Co(III) sets in at an earlier stage when the *para* isomer is reduced than is the case for the *meta*. The very facile internal electron transfer for the *ortho* isomer is worthy of note, but awaits explanation.

Other examples of using reducing agents to alter a functional group locally can easily be formulated, and the detailed study of processes of this kind can help to clarify problems of mechanism in the reduction process. A ligand can also be "damaged" merely by a change of the electronic energy producing a state in which internal electron transfer becomes possible, but this opportunity for investigating the properties of excited states produced photochemically has as yet not been exploited.

An extremely important question for the systems under discussion is whether, under any circumstances, an internal oxidant can assist an external oxidant in acting on a function located on a ligand. The internal oxidant Co(III) may be a special case because the orbital which engages the incoming electron has $\sigma$ symmetry with respect to the bond axis, and this does not overlap effectively with the $\pi$ orbitals of a ligand. When the acceptor orbital has $\pi$ symmetry, the effect of the external oxidant on a ligand such as 4-pyridinemethanol may be supposed to be felt at the internal oxidant center. A search for this kind of assistance by the internal oxidant was made by comparing the rates of oxidation of

$$\left[ (NH_3)_5CoN\underset{}{\bigcirc}-CH_2OH \right]^{3+} \quad \text{and} \quad \left[ (NH_3)_5RuN\underset{}{\bigcirc}-CH_2OH \right]^{3+}$$

under otherwise identical conditions. The redox potential for Ru(III) is very nearly the same as the "kinetic" redox potential for a cobaltammine, and, even though a net reduction of Ru(III) to Ru(II) cannot be observed (for reasons mentioned earlier), Ru(III) has the capacity to oxidize a carbon radical, and any assistance by the Ru(III) of the external oxidation should be reflected in a greater rate of reaction. The effect, if it exists, is very small. The values of $k$ ($\mu = 2.8$, $1.0$ $M$ $H^+$, $25°$), $\Delta H^{\ddagger}$, and $\Delta S^{\ddagger}$ for the Co(III) and Ru(III) complexes were found [7, 15] to be $1.3 \times 10^{-2}$, $24.7 \pm 0.7$, $15.3 \pm 2.0$ and $1.8 \times 10^{-2}$, $21.5 \pm 1$, $5 \pm 3$. The difference in rates at $25°$ is small, but there seems to be a real difference in the activation parameters. At this stage, it would be premature to attribute the difference to effects of the kind we are searching for, but it would be equally premature to give up the search at this stage.

## REFERENCES

1. A. Werner, *Z. Anorg. Chem.* **22**, 41 (1900); *Chem. Ber.* **49**, 896 (1911); *Am. Chem.* 386 (1919).
2. K. Schug, M. D. Gilmore, and L. A. Olson, *Inorg. Chem.* **6**, 2180 (1967).
3. S. M. Caldwell and A. R. Norris, *Inorg. Chem.* **7**, 1667 (1968).
4. P. Saffir and H. Taube, *J. Am. Chem. Soc.* **82**, 13 (1960).
5. R. T. M. Fraser and H. Taube, *J. Am. Chem. Soc.* **82**, 4152 (1960).
6. J. P. Candlin and J. Halpern, *J. Am. Chem. Soc.* **85**, 2518 (1963).
7. J. French, Ph.D. Thesis, Stanford University, Stanford, California (1968).
8. F. H. Westheimer, *Chem. Rev.* **45**, 419 (1949).
9. K. B. Wiberg, Oxidation by Chromic Acid and Chromyl Compounds, *in*

"Oxidation in Organic Chemistry" (K. B. Wiberg, ed.), Part A. Academic Press, New York, 1965.
10. R. Robson and H. Taube, *J. Am. Chem. Soc.* **89**, 6987 (1967).
11. F. Nordmeyer, University of Rochester, Rochester, New York, private communication.
12. E. S. Gould, *J. Am. Chem. Soc.* **87**, 4730 (1965).
13. G. Schweier, C. B. Storm, D. D. Thusius, and H. Taube, to be published.
14. E. S. Gould, *J. Am. Chem. Soc.* **88**, 2983 (1966).
15. D. P. Rudd, to be published.

# SUBJECT INDEX

## A

Acetate as bridging ligand, 69
Activation parameters for reductions
   of halo complexes, 59–62
Aluminum (III) (aquo ion)
   formula of, 10
   lability of, 10
Ammonia
   as blocking group, 19
   labilities of complexes of, 19, 20
   as ligand, 19
Antibonding orbital
   lability and, 22
   mechanism of electron transfer and,
     40
Aquo ions, determination of formulas
   of, 3–13
Azide ion
   in doubly bridged activated com-
     plex, 68
   reaction with HONO, 16

## B

Bridging groups, rate comparisons
   for, 51ff.

## C

Carbinol as oxidizable group, 80ff.
Cerium (IV)
   $vs.$ $Ce^{3+}$, redox potential, 4
   oxidation of ligands by, 79ff.
Chemical mechanism for electron
   transfer, 66, 67
Chloride ion as bridging ligand, 36,
   51ff.
Chromium (II)
   distortion of ion, 13, 22
   reactions
     with $Co(NH_3)_6{}^{3+}$, 28
     with $Co(NH_3)py^{3+}$, 28
     with cobaltammines, 51ff.
     with Cr(III) complexes, 44, 51,
       60
     with Fe(III) complexes, 60
     with Ru(III) complexes, 51, 60
     of trisbipyridine complex with
       Co(III), 32ff.
   reduction of ligands by, 63ff.
Chromium (III)
   ammine complex, 19
   aquo ion
     formula, 7, 8

lability, 4
reaction of $Cr(H_2O)_5O_2CCH_3^{2+}$
    with Cr(II), 44
    with V(II), 44
pentaammine complexes, reduction
    of, 51, 60
pentaaquo complexes, reduction of,
    44, 51, 58, 60, 63
Chromium (IV) as oxidant, 81, 91
Cobalt (II)
    absorption spectrum of, 5
    interchange rates for, 14
    low-spin, 90
    as product of induced electron
        transfer, 76ff.
Cobalt(III)
    ammines, labilities of, 19–20
    aquo ion, substitution lability of,
        4, 19, 24
    azidopentaamminecobalt (III),
        reaction with HONO, 16
    carbonatopentaamminecobalt (III),
        reaction with HONO, 17
    formatopentaamminecobalt(III),
        oxidation by $MnO_4^-$, 78, 92–94
    in induced electron transfer, 75
    interchange rates, 15, 16
    oxalatopentaamminecobalt (III),
        oxidation of, 77
    as oxidizing agent, 77, 81, 85
    pentaamminepyridinemethanol-
        cobalt (III), oxidation of, 80ff.
    reaction of aquopentaammine-
        cobalt (III) with $Cr^{2+}$, 36
        of $Co(NH_3)_5Cl^{2+}$ with $Cr^{2+}$,
            36–37
        with $Cr(bipy)_3^{2+}$ 32, 34,
        of hexaamminecobalt (III) with
            $Cr(bipy)_3^{2+}$, 32
        of pentaamminecobalt(III) with
            $Cr^{2+}$, 51ff, 60
        with $V^{2+}$, 44

Coordination number
    absorption spectrum and, 5, 6
    determination of, 3–13
Coordination sphere
    changes on electron transfer, 1, 2
    distortion of, 3, 6, 13
    labilities, 3, 4
Cu(II), aquo ion
    distortion of, 13
    lability of, 4
    NMR data on, 13, 22

D

Doubly-bridged activated complex,
    68–69

E

Electron transfer
    induced, illustration, 75.
    internal, 74.
    mechanisms of, 47, 50, 62–66
    reducibility of ligand in, 64
    role of vacant orbitals in, 50
Electron tunneling, 66
Electronic structure
    lability and, 21, 22, 24
    mechanism of electron transfer and,
        29, 38, 67
Entropy of activation as diagnostic
    of mechanism of electron transfer,
    59–66
Equilibria for activated complexes,
    53–57
Europium (II) as reducing agent, 60
Excited electronic state in electron
    transfer, 34

F

Fluoride ion as bridging group, 63
Formyl group, oxidation of, 82.
Fumarate as bridging ligand, 68.

## G

Glycolate as bridging ligand, 65
Glyoxylate as bridging ligand, 65

## H

Halide ions
    as bridging ligands, 51ff., 59ff., 69
    effect on rate of reduction by V(II), 57
Hydroxide ion
    as bridging ligand, 37
    as ligand in outer-sphere mechanism, 34, 35, 43

## I

Induced electron transfer (illustration), 75
Inner-sphere mechanism for electron transfer
    changes in electronic structure and, 34
    effect of changes in dimensions of reactants, 33–34
        of counter ions, 30, 31
        of ligands, 31–35
    electron delocalization and, 33–34
    isotopic effects, 34
    Marcus correlation, 30
    salt effects, 29–31
    solvent effects, 35
Interchange rates
    dipositive ions, 14
    tripositive ions, 16
Intermediates
    in induced electron transfer, 76ff.
    pentacoordinated, 18–19
        as products of reduction of Co(III), 90
Internal electron transfer
    induced, 73–74

    as probe of mechanism, 81–82
    rate comparisons, 87–88
Ion solvation, general, 23–24
Iron(II) (aquo ion)
    interchange rates, 14
    as reducing agent, 60
    substitution lability of, 4
Iron(III) (aquo ion)
    substitution on, rate comparison, 18, 19
    substitution lability of, 4, 18
Isonicotinamide, as ligand, 67
Isotopic dilution, 6–10
Isotopic fractionation, 8, 10, 40

## K

Kinetic isotope effect (H vs. D), 34, 37, 89

## M

Marcus correlation, 30, 35

## N

Net activation process, 55
Nickel(II) (aquo ion)
    interchange rates, 14
    monoammine complex, 19
Nitrobenzoate, reduction of, as ligand, 95
Nuclear magnetic resonance
    coordination number and, 10, 11
    exchange rate and, 11–14

## O

$^{17}$O,NMR studies
    chemical shift, 11
    line-broadening vs. temp., 11
    molal shift, 11

$^{18}O$
  in exchange of water, 6–10
  isotopic fractionation, 8, 10, 40
  in oxidation of formate by $MnO_4^-$, 93
Orbital symmetry
  electron transfer and, 38, 41, 58–59, 67–68
  induced, 75, 96
Outer-sphere activated complex
  definition, 28
  isotopic effects, 34
  ligand effects, 31–35
  salt effects, 29–31
Oxalate ion in induced electron transfer, 77

**P**

Pendent groups in electron transfer, 69
Permanganate ion
  exchange with $MnO_4^{2-}$, 31
  oxidation of formate as ligand by, 78
Permeability for electron flow, 57
Pseudorotation, 13, 22

**R**

Radical intermediates, 87ff.
Rate comparisons
  in substitution, 13ff.
  of oxidation of ligands, 79
Rates of electron transfer
  bond distortion and, 33
  conjugation and, 67
  electronic structure and, 67
  reducibility and, 64ff.
  interrelation with substitution, 42
Redox potentials, 4
Reducibility and rate of electron transfer, 64–67

Ruthenium(II), substitution lability of, 4
Ruthenium(III)
  hexaammine complexes
    in induced electron transfer, 76, 96, 97
    self-exchange, 33
  pentaammine complexes, reductions of, 51, 60
  trisethylenediamine complexes
    self-exchange, 33
  substitution labilities of, 4

**S**

Self-exchange
  $Fe^{2+}$—$Fe^{3+}$, 27
  $Fe(CN)_6^{4-}$—$Fe(CN)_6^{3-}$, 29
  ligand effects in, 70
  rate and bond distance, 33
  $Ru(NH_3)_6^{2+}$—$Ru(NH_3)_6^{3+}$, 33
  $Ruen_3^{2-}$—$Ruen_3^{3+}$, 33
Silver(II), catalyst in induced electron transfer, 81, 82, 84–86
Substitution
  anation, 17
  intermediate in, 17
  mechanism of, 15, 17
  product competition ratio in, 16–18
Substitution *rate*
  comparison with water exchange rate, 4, 13, 19, 20
  effect of charge on, 20, 22
    of electronic structure on, 20, 21
  excited electronic states and, 24
  structure and, 22

**T**

Thiocyanate ion as ligand, 17, 18, 44
  oxidation of, 95
Two-electron oxidation, 41, 45, 75, 77, 81, 93

## V

Vanadium(II) (aquo ion)
  reaction with hexaamminecobalt
    (III), 57
  interchange rates, 14
  oxidation by facial attack, 44
    with pentaamminecobalt(III),
      60ff.
  as reducing agent, 44

substitution lability of, 4
Vanadium(III) (aquo ion)
  mechanism of substitution, 19, 21
  substitution lability, 4, 18

## W

Water as ligand in electron transfer,
  34–36, 43